URETHANES IN COATED FABRICS
VOLUME 2

(Major Papers from the
Journal of Coated Fabrics)

TECHNOMIC
PUBLISHING

a TECHNOMIC® publication
TECHNOMIC Publishing Co., Inc.
265 Post Road West, Westport, CT 06880

URETHANES IN COATED FABRICS
VOLUME 2

(Major Papers from the
Journal of Coated Fabrics)

265 Post Road West, Westport, CT 06880

a TECHNOMIC® publication

Printed in the U.S.A.
LCC 73-90329
ISBN No. 087762-279-5

CONTENTS

PUBLISHER'S NOTE

*All of the material in this publication has been reprinted from copyrighted articles in the **Journal of Coated Fabrics**, 1974, 1975, and 1976. The copyright owner is Technomic Publishing Co., Inc., Westport, Connecticut.*

J. GILLIBRAND

J. B. Broadley Ltd.
Rossendale, Lancashire, England

POLYURETHANE COATED FABRICS

(Received November 5, 1973)

ABSTRACT: A discussion of the factors which have produced the remarkable growth in the use of polyurethane-coated fabrics, and of the factors which influence their properties.

The latter includes a description of the various techniques used in the manufacture of PU-coated fabrics and of the way in which variations in manufacturing techniques can affect the finished products.

This is integrated with a description of the basic properties of polyurethane-coated fabrics such as scuff resistance, flexing endurance, last fastness and permeability.

THE RECENT GROWTH in importance of polyurethane-coated fabrics has been quite phenomenal. A few years ago they were used for certain applications in Europe, and were talked about in Britain and the United States. Today, though it is difficult to quote exact figures, many millions of yards are used annually for consumer goods, such as clothing, footwear, upholstery and handbags.

There has to be a reason for this growth, and it lies in the particular technical and aesthetic properties of polyurethane-coated fabrics. Technically, these materials are superior to vinyl-coated fabrics in a number of important ways. Firstly, polyurethanes can form soft, flexible films without the aid of plasticisers. This means that there are no oils to be washed out by solvents, and therefore the coatings can be formulated to be dry cleanable. The absence of plasticisers also means that there are no migration problems where the coated fabrics are used in conjunction with other materials, for example with adhesives or leathers. Secondly, polyurethanes are much more resistant than PVC to stiffening at low temperatures. Thus a garment made from polyurethane will still feel supple on a cold day in winter, whereas a PVC garment would feel stiff and uncomfortable. Also the PU-coated fabric is more resistant to flex cracking at low temperatures, an important property for garments and footwear, and for car upholstery. Thirdly, polyurethane is very resistant to abrasion in thin films, enabling lightweight, but serviceable coatings to be produced. Fourthly, in thin films polyurethane has a high permeability to water vapour, yet is waterproof. Advantage can be taken of this, in conjunction with manufacturing and structural techniques, to produce PU-coated fabrics with a measure of permeability.

Again, aesthetically, PU-coated fabrics offer many advantages over vinyls. The softness and suppleness of PU coatings result in excellent drape properties for clothing, properties which can be enhanced by the selection of flexible base fabrics, such as warp knitted nylon or circular knitted cotton. By selection of polymer and compounding ingredients, PU's can be formulated to give a very dry handle, even in high gloss films, or they can be given the somewhat greasy handle characteristic of certain leathers. By virtue of their construction they have a pleasant, characteristic leather "break," and this same construction imparts a softness and fullness to the touch.

This remarkable combination of aesthetic and technical properties accounts for the growth in usage of PU-coated fabrics. To date, generally, this has not been growth at the expense of vinyl-coated fabrics, though there are certain fields in which this could happen in the future. The polyurethanes have provided an additional and novel coated fabric, which has opened up new markets, and has provided further opportunities in those markets approaching saturation.

However, with all pros there are cons, and in the case of polyurethane, there are two distinct disadvantages with respect to vinyl. Firstly, it is much more expensive, the actual cost of raw material in the coating being from 10 to as much as 15 times more expensive than a vinyl coating of equivalent thickness. Secondly polyurethane must be applied from a solvent-based coating compound. Vinyls are applied from a paste which is virtually 100% solids. For example, if a vinyl paste coating having a thickness of ten thousandths of an inch is applied to a substrate, after the heating operation a coating thickness of approximately ten thousandths of an inch is obtained. With polyurethanes, where the polymer is dissolved in solvent, the solution may have a solids content of perhaps 25%. Thus on spreading a polyurethane solution of ten thousandths of an inch onto a substrate, after drying off the solvent, one is left with only 2.5 thousandths of an inch coating thickness.

This factor of solution coating, plus the high cost of polyurethane, means that the production of thick PU coatings is a laborious and expensive operation. Thus PU coatings tend to be thin. From a coated fabric viewpoint, this immediately imposes certain restrictions. Fabrics are not flat surfaces. They have a pattern imposed by the weave, and they have frequent, usually small weaving or knitting imperfections. A thin coating applied to the fabric will not hide this surface pattern as a thick coating would. This is illustrated diagrammatically in Figure 1.

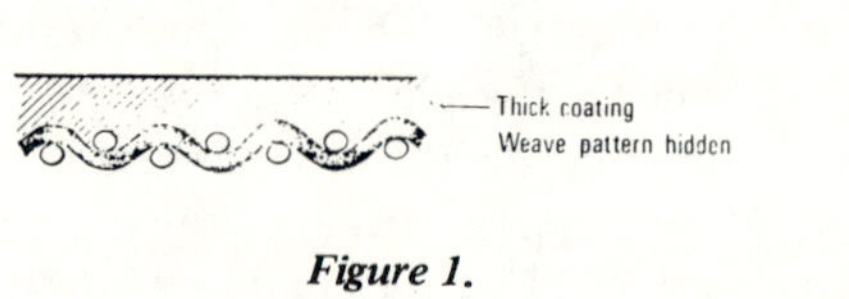

Figure 1.

If I appear to be making a lot of this simple matter, I would ask your patience, because this problem is a fundamental one of PU-coated fabrics, and the method adopted to overcome it is responsible for much of the aesthetic appeal of these materials, and also for many of their faults. The method adopted is to provide a cushion between the basic fabric structure and the coating. This is done in one of two ways: either the fabric is raised by a brushing technique and the coating is applied to the tips of the fibres, or a polyurethane foam of perhaps a half millimetre thickness is adhesive or flame bonded to the fabric, and the coating applied to the foam. Both of these systems are drawn in Figure 2.[1]

This cushion of either fibre or foam (and raised fibre is by far the commonest method) is responsible for the fullness and softness of PU-coated fabrics, and for their ability to produce a leather-like break as the coating folds into the cushion. It is also responsible for the three major failings of these materials: poor scuff resistance, delamination of coating, and failure under flex and stress conditions. All these I will discuss in more detail at a later stage, but now I would like to deal with manufacturing techniques, these being very relevant to any discussion of properties.

There are two basic methods of producing PU-coated fabrics, as indeed there are with vinyls, namely direct and transfer coating, but the problems involved are quite different from those of vinyl-coated fabrics.

First let us take the case of direct coated products. The starting substrate is a raised pile fabric, for the reasons already discussed and the objective is to lay the coating on top of the pile, with sufficient penetration to achieve good coating adhesion, but not so much as to saturate the pile. The fabric is all important. For optimum results it should have a directionally laid, even pile, as indicated in Figure 3a.

An evenly cropped pile as indicated in 3b would lead to a product of rippled, uneven appearance, while a randomly laid pile as indicated in 3c would yield a product with too much coating penetration, and a poor handle.

In the previous drawings, for simplicity the fabrics are shown with the pile raised from both warp and weft yarns. In fact the pile is almost entirely raised from the weft yarns, and the fabrics are specially constructed to enable this to be done.

Figure 4 indicates two methods of achieving this. The first uses a relatively low twist yarn and a twill construction. The drawing shows a 3–1 twill, but this could also be a 2–1 or a 4–1 twill if desired. The twill construction presents a large area of weft for raising and enables a smooth pile to be obtained. The second method uses a very low twist condenser weft in a plain weave construction. The low twist yarn is very bulky compared to the warp, and quite a good raising can be obtained.

The actual coating operation is carried out by means of a knife-over-blanket coating head, as drawn in Figure 5.

---

[1] True in 1971. Now a third method available, namely the use of chemically expanded PU or a foamed PU emulsion as tie coat.

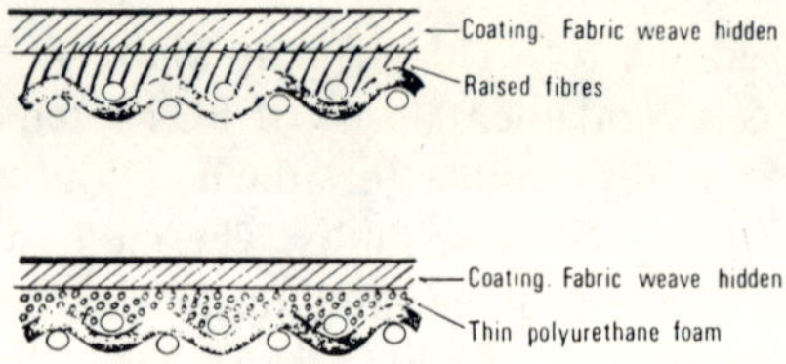

Figure 2.

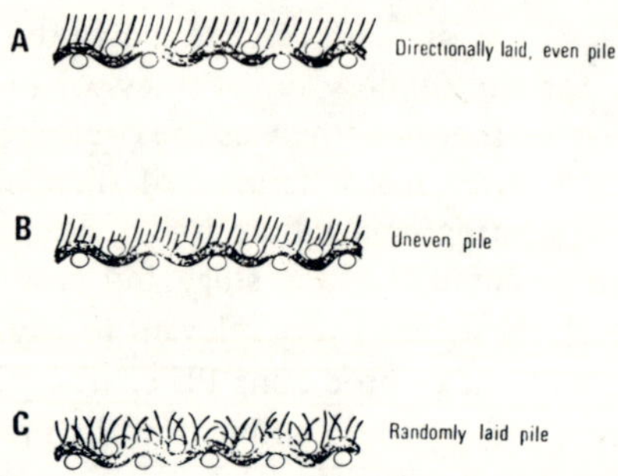

Figure 3.

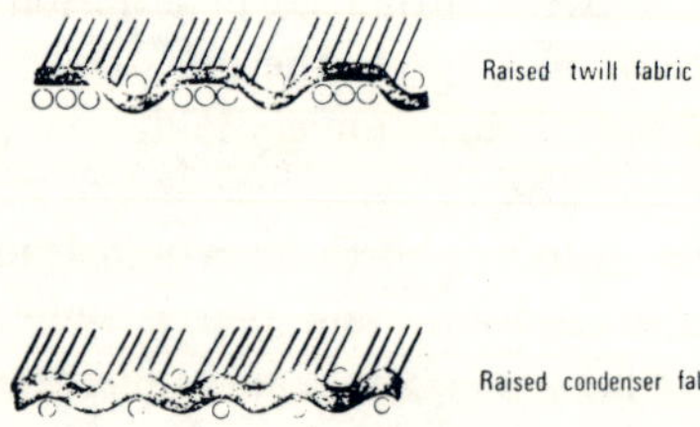

Figure 4.

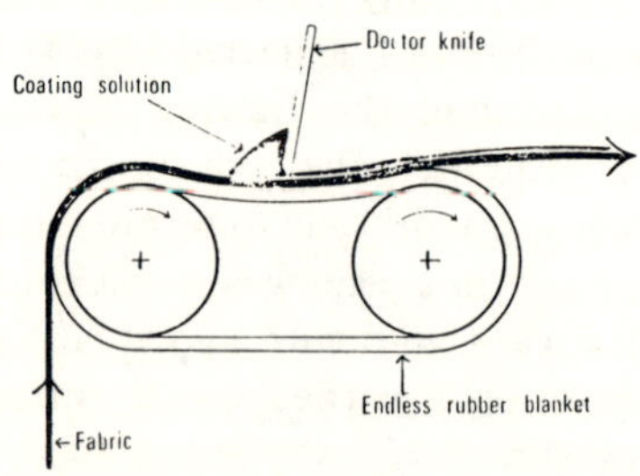

Figure 5. Blanket coating head.

The fabric, lined up so that the pile is directed away from the doctor knife, is carried under the doctor knife by means of an endless rubber belt or blanket, and the pressure of the doctor knife against the fabric, the shape of the doctor knife, the viscosity of the coating solution, and the speed of running are all critical in determining the end product. However, no matter how well controlled are these factors the fabric must be initially of good quality.

Machinery for carrying out the operation is shown in Figure 6.

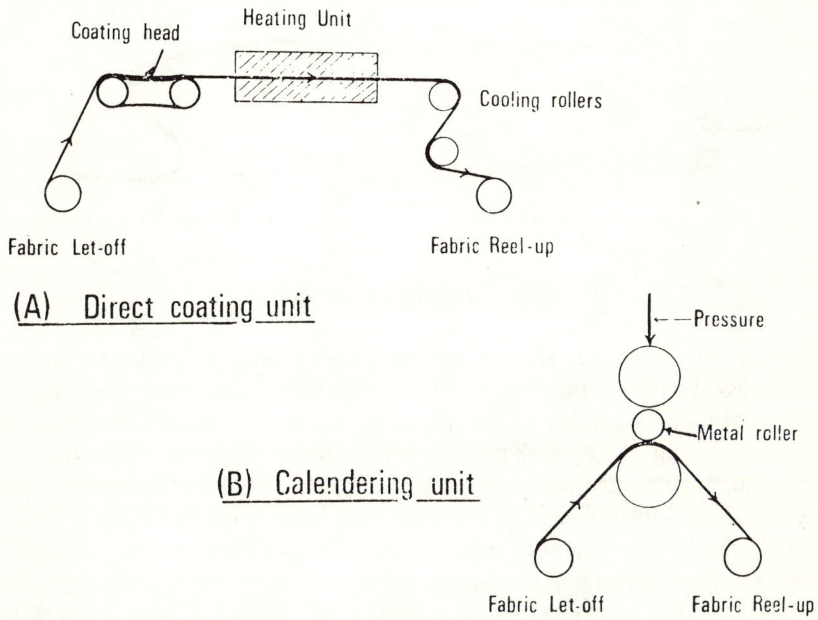

Figure 6.

Figure 6a indicates a blanket coating head in line with a heating unit to dry off solvent, and this may be an oven or a hot plate system, followed by a cooling unit to ensure that the coating is cold before reeling. This is important, otherwise the whole product may stick to itself in the reel. Figure 6b indicates a simple calendering unit, this being the process of heavy rolling to which the coated fabric is subjected after one or two initial coats. This evens out the products before further coats are applied.

The whole process of direct coating is a highly critical operation, and is utterly dependent on the condition of the substrate. For many years the development of this operation was held up in Britain because of the absence of suitable fabrics. Indeed, good quality pile fabrics are still not produced in this country. Fortunately, another coating method has been developed which enables fabrics with an inferior degree of raising to give acceptable results. This is the transfer coating system, which also has the advantage of enabling stretchy fabrics, such as warp knitteds, to be coated. Figure 7 shows an example of a transfer coating line-up.

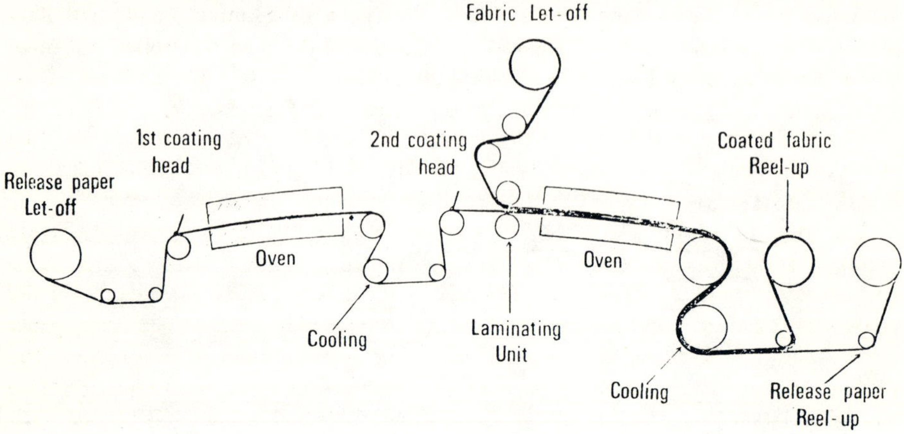

Figure 7. Transfer coating line-up.

There are two coating heads and two oven drying units. A silicone-treated release paper is passed through the first coating head, and a "fixed gap" coating of polyurethane solution is applied to the paper by means of a knife-over-roller coating unit. This first coat, the "top skin" is then dried in the first oven unit, cooled, and a second coat, "the laminating coat," applied at the second knife-over-roller coating unit. The fabric is then pressed, raised side down, onto this wet coating by means of a roller system. The whole then passes into the second oven unit, where the solvent is removed. Then depending on the type of coating used, the coated fabric is either separated from the release paper, as shown in the diagram, or is reeled with the release paper for separation at a later stage. If the latter, it may be necessary to use a paper with a release coating on each side. In this particular coating line, the paper follows an arched path through the machine. This assists handling of the paper, but some people may argue that arching is unnecessary.

The whole process sounds delightfully simple, but its practical application is a good deal more complex than it seems. For instance, the speed of running and the oven temperatures must be balanced to give coatings free from blisters or retained solvent. The release paper must be selected to match the polymer system used. The pressure used at the laminating nip is critical in determining the properties of the finished product. The solvent system used for the laminating coat must be carefully selected to avoid wrinkling or crazing of the top skin. One could talk about these problems for hours, and in fact, the industry has talked about them for years.

The selection of release paper can determine the degree of sheen of the product. For example, high gloss release papers have been used to produce large quantities of "wet-look" coated fabrics. Choice of a matt paper will correspondingly yield a matt appearance, and so on. Embossed release papers can be used to put a grain into the coating, though because of the thinness of the coatings, there are some limitations to effective grainings. Deeper grains show up quite well, but lighter, kid type grains tend to be lost in the leather-like "break" of the PU-coated fabric. This "break," produced by crumpling the material, is the obvious way to put a graining into a PU-coated fabric. It can be done on large scale by tumbling or some similar operation.

It is possible to post emboss the coatings, though this requires heat to be at all effective, and the hot embossing of coated pile fabrics can have a drastic effect on handle. We have never found this to particularly enhance the aesthetic appeal.

The printing of PU-coated fabrics can be carried out by conventional gravure systems, using polyurethane-based inks. To give high protection to the ink, or to give an appearance of depth, the printed coated fabric can be again doctor knife coated with a clear polyurethane solution, or alternatively a clear top skin on paper can be printed, followed by a second top skin and the lamination process. The process of doctor knife coating polyurethane solutions onto coated fabric has also been used to produce "wet-look" materials.

There is another possible method of producing PU-coated fabrics, and this is by the lamination of polyurethane film to the fabric. This method which could be suitable for the production of large quantities of a single colour, lacks the flexibility of the transfer coating system, which allows easy selection of compounds and laminating pressures.

The chemistry of polyurethanes is an extremely complex subject, very well covered in a range of literature. For this discussion of PU-coated fabrics I would like to reduce any chemistry to its simplest possible terms, and run the risk of having this called oversimplification by the specialist. Basically there are two types of polyurethane used: two-component systems where isocyanates are added to polyol systems shortly before coating, and which produce crosslinked non-thermoplastic coatings, after the necessary curing operations, and single component systems which do not require a curing operation, and which produce heat sensitive, and therefore weldable coatings. The single component systems have a higher molecular weight than the two component systems, this being necessary to develop adequate physical properties, and consequently they need more efficient solvents to dissolve them. Single components use dimethyl formamide and tetrahydrofuran, whereas two components may use ethyl acetate.

PU-coated fabrics can be produced entirely from two-component systems, and such systems with a choice of polyol and type and amount of isocyanate can make available a wide selection of drape properties. These coatings also have better resistance to high rate, frictional abrasion. However, they suffer from the disadvantage that in the transfer coating process they must be left on the release paper for a period of time to allow cure to take place. This means that one cannot instantly see the product being made at the end of the machine, and so have the opportunity to correct faults.

The single component systems do not offer the same ability to select drape properties[2], but only require to have the solvent removed to leave a film with fully developed properties. Thus, these can be separated from the release paper immediately after making. Also the solutions have a virtually unlimited pot life.

In practice it is common to use a single component system for the "top skin," enabling instant release, and a two-component system for the "laminating coat," giving a wide choice of drape properties. Crosslinkable acrylic latices may also be used as laminating coats, but those, unlike two-component polyurethanes, will not normally cure at room temperatures, and so require greater heating during processing.

Whichever system is used, colour matching and colour reproducibility in PU-coated fabrics is an aggravating problem. The basic difficulty is that the coatings are very thin, and high pigment concentration can cause a deterioration of flex properties. Thus the coatings tend to have low opacity, and their colour is greatly influenced by the colour of fabric substrate. This substrate must be obtained to match the colour of the coating as nearly as possible, but even so colour reproducibility can be difficult, because even the pressure by which the fabric is applied to the coating can affect the shade. Very close control of fabric colour, and all coating operations is necessary to obtain reproducible results.

Having dealt briefly with the manufacturing methods used to produce PU-coated fabrics, and with some of the problems encountered, I would now like to revert to a discussion of some of the properties of these materials, with particular regard for some of the problem areas. I would list these as follows:

Scuff resistance:	coating adhesion:	flex resistance:
Hydrolytic resistance:	permeability:	light fastness

By virtue of their structural make-up PU-coated fabrics are highly susceptible to scuff damage, by which I mean impact against sharp or rough surfaces. The coatings are thin and supported only by a fibre cushion, and are consequently easily pierced. Provided that the coating adhesion is good, the damage will be only local, but if the coating adhesion is poor, the scuff damage can be a focal point for separation of whole areas of the coating. This was commonly seen in some of the earlier PU-coated fabrics used, for example, in handbags. Obviously the heavier the coating, the greater will be the resistance to scuff damage. Some of the materials we produce for footwear with a coating weight of approximately 110 gm per sq metre, have moderately good scuff resistance, certainly they are sufficiently durable for fashion shoes. Other materials, for example for light-weight garment applications, may have only 50 to 60 gm per sq metre, and therefore much lower scuff resistance. It is more logical to use coating weights with reference to PU-coated fabrics then to use thickness, for reasons which will become apparent as the discussion continues.

The adhesion of the coating to the fabric depends upon a number of factors. Clearly it must depend upon the pressure exerted in laying the fabric against the

[2] No longer true.

wet laminating coat, and also upon the thickness of the laminating coat. These two factors govern the penetration of coating into the raised pile, and bond strengths ranging from one or two pounds up to seven or eight pounds per inch width can be obtained by altering these factors. The latter figure is probably approaching the maximum consistently obtainable in transfer coating, though figures up to nine or ten pounds can be obtained by direct coating an adequately raised fabric. The lamination process becomes more critical as the quality of the raised pile deteriorates. Figure 8 indicates the reason for this.

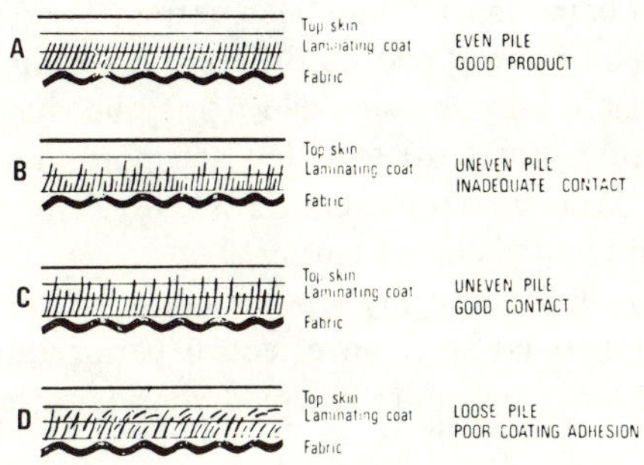

Figure 8.

In Figure 8a the pile is evenly cropped and is well gripped by the laminating coat. In Figure 8b the pile is uneven, and the laminating coat has not made adequate contact. In Figure 8c the pile is again uneven, but more pressure has been applied by the laminating roller, forcing adequate contact. Also the pile of the fabric must be free from loose fibre, or this can cause loss of adhesion as shown in exaggerated form in Figure 8d. This is true of either direct or transfer coated products.

The flex life of polyurethane films is normally quite good, though certain polymers do produce harder films than others, and some of these do have a poor flex life. Given flexible polymers, PU-coated fabrics can still have a poor flex resistance if blisters are allowed to form in the coating during processing. This can happen if oven temperatures are too high, causing surface drying of the solvent-based coating, and trapping solvent within the coating. This trapped solvent forms blisters on further heating, and these blisters are potential points of breakdown during flexing operations.

Also the flex resistance can be worsened if the pile of the fabric is allowed to penetrate into the top skin of the coated fabric. This can happen during the lamination process. The laminating coat softens the top skin, and if heavy pressure is used in lamination, the fibres can be pushed into the top skin, and can actually penetrate to the surface. This is particularly true of loop raised, warp knitted nylon fabrics. The nylon loops can be relatively rigid, and easily pushed through the softened top skin. Thus, the finished product has a surface literally chopped up by nylon loops, and on flexing, or stretching, or both, the coating falls apart. This penetration can be avoided by choice of correct solvent balance for the laminating coat, and by avoidance of excessive pressure.

Regarding hydrolytic resistance of polyurethanes, much valuable work has been done, particularly at SATRA. At the time of compiling this talk, however, I am very far from being convinced of the importance of hydrolysis as a failure factor in PU-coated fabrics, as opposed to poromerics. All the examples of so-called hydrolytic failure which I have seen in PU-coated fabrics could have had a much simpler explanation, such as inadequate coating adhesion, blisters, or fibre penetration. Also test results on PU-coated fabrics have been extremely erratic, the same samples being classed as very good or very bad on different occasions. I would like to keep as open a mind as possible on this issue, and it may be that a clearer picture will emerge in the near future, but I do feel that the poromerics, with their heavier microcellular coatings, ideal sites for chemical reaction, are more prone to hydrolytic failure than the simpler PU-coated fabrics.

Water vapour permeability is a highly relative subject. What the Shirley Institute, Manchester, England, might consider an effective permeability level for garments can be quite different from what SATRA would consider effective for footwear, or FIRA[3] for upholstery.

How permeable are PU-coated fabrics? Generally speaking their permeability lies below 1 to 1.5 mgm/sq cm/hr, and can be as low as 0.2 when heavy coatings are used, or when measures are taken to ensure that the coating is free from blisters, pin-holes, or fibre penetration. Thus the factors which lead to higher permeability levels can cause trouble with other physical properties. To achieve both, a compromise is necessary.

The light fastness of PU-coatings has given justifiable cause for concern. The polyurethanes produced from aliphatic or alicyclic isocyanates have better light fastness than those produced from aromatic isocyanates, and though some single component polyurethanes have excellent light fastness, these tend also to form more crystalline, less flexible coatings. From a colour standpoint, polyether polyurethanes have good light fastness, but unfortunately are embrittled by an oxidation process on exposure to light. Ultra-violet light absorbing agents, fillers and pigments obviously help, but do not solve the problem. In fact, currently, there is no single, simple answer, and compromise again is a necessary requirement.

We have, at this conference, very competent speakers on the subjects of artificial leathers for footwear, clothing, upholstery and the handbag industries, and I am sure that they will have something to say about PU-coated fabrics and the advantages and disadvantages within their own particular fields. Consequently, I will not waste time by overlapping their comments. There is, however, one particular type of PU-coated fabric which has caused a great deal of trouble over the past year or so, sufficient trouble indeed to put the whole of these materials in a bad light, and I

[3] Furniture Industry Research Association

would like to make some comments on this material. I am referring, of course, to the PU-coated loop raised, warp knitted nylon used for ladies' stretch boots. This material is required to have very high stretch in one direction, and very low stretch in the other. The high stretch is used in the girth of the boot and the low stretch in the length. The material is subjected to a very high distortion as the foot is thrust through the narrow ankle of the boot. Given a good, flexible coating, a base fabric of adequate strength, and sensible reinforcing, it is just possible for the material to survive this application. However, so many inferior materials have been used, materials with blistered and fibre penetrated surfaces, materials with inadequately cured acrylic laminating coats, materials with brittle coatings, that there has been widespread failure, and consumers must be very wary of these products in the future. It is a great pity that the public has been subjected to this misuse of badly made material.

In closing, it is obvious that PU-coated fabrics have come a long way in a short time. In my view they are still in an early development stage, and there is much more to come. With close co-operation between user and producer, problems like those with the stretch boots can be overcome, and these materials will have an assured future.

J. S. Maddams MA

Head of Test Development Department
Shoe and Allied Trades Research Assoc. (SATRA)
Kettering, Northants, England

A REVIEW OF THE PROPERTIES AND TESTING OF PU-COATED FABRICS

(Received December 1, 1973)

KEY WORDS: Urethane fabrics, shoes, leather.

INTRODUCTION

The properties of PU-coated fabrics (PUCFs) which affect their shoemaking behaviour and wear performance are to a large extent determined by their special structure. Although they are seemingly fairly similar to PVC-coated fabrics there are, in fact, differences in structure and consequently a number of important differences in properties.

The need for a well-raised fabric to which the thin polyurethane coating is applied dictates the construction of the base fabric. In such fabrics it is always the weft (across) fibres which are raised, so one surface must be nearly all weft yarns. This is achieved by the use of a 4 × 1 twill fabric, but in addition the weft yarns are made heavier than the warp (along) yarns and the crimp in the warp yarns is reduced to a minimum so that they do not appear at the surface which is to be raised. These last two features of the base fabric produce important characteristics in the properties of the whole material.

GENERAL STRENGTH AND STRETCH BEHAVIOUR

The strength and stretch of the coated fabric determines how it should be treated when lasting the upper. The way the upper is stretched in lasting is complex but important differences in behaviour between PUCFs, PVC-coated fabrics and leather are revealed by a comparison of their strength and stretch properties ('along' and 'across' the fabric roll or skin). The values given in Table 1 are averages provided by the records for upper materials in the SATRA Data Bank. Nearly all the results for the three types of coated fabrics are for women's weight materials so they have been compared with a women's weight leather of 1 mm thickness.

Table 1 shows that solid PVC-coated fabrics are only a little lower in breaking load than leather but cellular PVC-coated fabrics are somewhat weaker still. The

Table 1. Strength and Stretch Properties (Average Figures)

	Breaking Load (kgf/cm)		*Breaking Extension (%)*	
	Along (Warp)	*Across (Weft)*	*Along (Warp)*	*Across (Weft)*
Leather (1 mm)	19.6	17.6	36	56
Solid PVC on woven fabric	19.2	15.0	10.0	19.0
Cellular PVC on woven fabric	15.9	12.9	12.0	18.0
PU-coated woven fabric	10.3	13.1	8.0	22.0

weft direction of PUCFs is comparable in strength with the weft direction of the other fabrics. The warp direction, however, is considerably weaker than the warp direction of the other fabrics. It is in fact usual, as these figures show, for the warp of a PUCF to be weaker than the weft, whereas with other fabrics it is usual for the warp to be stronger. The comparatively light weight of the warp yarns in PUCFs is the reason for their low strength.

Table 1 also shows that the values of breaking extensions for all coated fabrics are considerably less than the values for leather. It is usual for the weft yarns in a fabric to be more stretchy than the warp but the difference is particularly marked for PUCFs due to the small amount of crimp in the warp yarns and the increased amount in the weft yarns.

It is thus evident that the warp yarns of a PUCF are weak and of low breaking extension. For shoemaking the practical implications are:

a. Reinforcement of the material with a backing fabric is a wise precaution.
b. The stretch applied in lasting should be kept to a minimum.
c. Vamps should be cut with the warp yarns in the heel/toe direction because the overall stretch in this direction during lasting is least. Actually the stretch in the extreme toe region is fairly high but it is likely that the tension in the warp yarns here and further back in the vamp is distributed uniformly by internal yarn slippage.

STRENGTH AT A STITCH LINE

The strength of an upper material at a row of stitching close to a cut edge is measured by SATRA Test Method PM33, using the apparatus shown in Figure 1. A row of size 14 round point machine sewing needles is inserted through the material at a fixed small distance from a cut edge so as to correspond to a row of stitches. An increasing force is then applied to the material at right angles to this row until failure occurs. With leather it is usual for the failure to be tearing along the row of holes (as with a postage stamp), although unless the leather is thin the failure loads *are higher than are experienced in practice.* With fabrics and coated fabrics the yarns between the "line of stitches" and the edge of the material are pulled off the yarns at right angles to them.

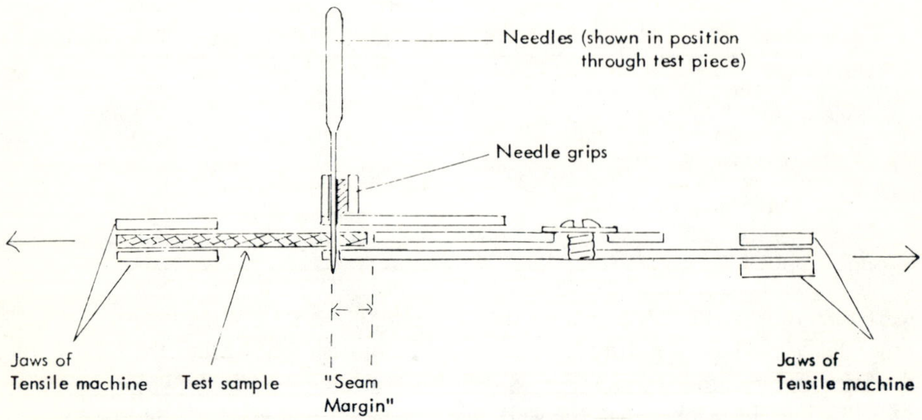

Figure 1. The SATRA attachment for use with most tensile testing machines which permits the seam strength of the material to be assessed (SATRA Test Method PM33).

Average values for failure of the various types of coated fabrics are given in Table 2.

Table 2. Strength at Stitch Perforations (Average Figures)

	Failure Load (kgf/2.5 cm width)	
	Along (Warp)	***Across (Weft)***
Solid PVC on woven fabric	32.9	25.4
Cellular PVC on woven fabric	28.4	24.8
PU-coated woven fabric	5.2	25.9

The one striking result in Table 2 is the very low value for PUCFs when the applied force is in the warp direction, i.e., when the weft yarns are pulled off the warp. This is largely because of the small amount of crimp in the warp yarns, so that it is easy for the weft yarns to slide off them. (In the other direction of test the crimp in the weft yarns prevents the warp yarns from pulling off easily.) Consequently it would be unwise to make a seam in the weft direction with the edge of the material left raw or unreinforced.

TEAR STRENGTH

Comparative figures for the tear resistance of leather and coated fabrics using the tongue tear test (SATRA Method PM30) are given in Table 3.

Table 3. Tear Resistance (SATRA Method PM30)

	Load to Tear (kgf)	
	Along (Warp)	*Across (Weft)*
Leather (1 mm)	2.8	
Solid PVC on woven fabric	3.6	3.4
Cellular PVC on woven fabric	2.8	2.8
PU-coated woven fabric	3.6	2.8

It will be seen that in terms of load there is little difference between the various materials. The PUCF, however, usually shows a different type of failure when the tearing force is in the weft direction. Instead of the warp yarns tearing, the weft slip off of them, in the same way as in the seam strength test. If the test-piece size is increased so as to improve the grip between warp and weft yarns the failure changes to breakage of the warp threads at a somewhat higher load. The comparable tearing loads for PU- and PVC-coated fabrics (even though the breaking load of the PUCF is lower than that of the PVC-coated fabric) are probably the result of the twill weave of the former and the looser attachment of the coating to the base fabric. Both of these features would produce a wider and more even distribution of tearing stresses and hence a higher tearing load.

The general conclusion from the laboratory tear tests of PUCFs is that they should not be especially susceptible to this type of damage in shoe manufacture or in wear.

LATERAL CONTRACTION IN LASTING

Lasting a shoe upper usually involves some contraction of the material near the feather edge of the last in the direction of the featherline. This contraction is achieved in two quite different ways:

a. By the flattening of pleats in the upper at the featherline by the wiper plates.

This effectively produces an edgewise compression of the material.

b. By pulling the upper at right angles to the featherline so that it contracts in the direction parallel to the featherline (the Poisson's ratio effect).

Examination of PUCF shoes shows that optimum contraction at the toe is often not achieved because here the upper shows many fine pleats at right angles to the feather edge.

The SATRA Wrinkleometer (Test Method PM44) provides a way of measuring the edgewise compression behaviour of upper materials and shows that the PUCFs are not worse than other synthetic materials of leather. This therefore suggests that the pleating at the toe is due to insufficient lateral contraction, during lasting extension.

With plastic-coated fabrics the lateral contraction due to extension is mainly due to straightening of the yarns being tensioned.

This increases the crimp in the yarns at right angles so that the fabric must contract in this direction. The warp direction of a PUCF, which is the one at right angles to the featherline at the toe, has little crimp, however, so it can produce little increase in the crimp of the weft yarns when it is stretched, and hence little lateral contraction. At the sides of the forepart, where lasting stretches the weft yarns, an appreciable contraction of the warp yarns is possible so that wrinkling there is not to be expected.

MOIST HEAT SETTING

Typical results for the moist heat setting properties of coated fabrics and leather using SATRA Test Method PM18, are shown in Table 4. The setting conditions are 1½ minutes in steam at 60°C followed immediately by 3½ minutes at 120°C in a hot air stream.

Table 4. Moist Heat Setting (SATRA Test Method PM18)

	Percentage Set After One Week's Recovery
Chrome leather	72
Solid PVC on woven fabric	70
Cellular PVC on woven fabric	69.5
PU-coated woven fabric	56

PVC-coated fabrics and chrome leather are closely similar in their set retention but PU-coated fabrics are appreciably worse; however, the amount of fall-in in practice is also dependent on the amount of stretch applied in lasting. If, as is likely, the amount of stretch of a PUCF is less than that of a leather upper, then the visible amount of fall-in may be no more, even though the percentage set is appreciably lower.

Where some improvement in the set of PUCFs is desired, this may be achieved by the use of dry heat setting at a higher temperature and for a longer time.

COATING ADHESION

Since the PU coating is applied to a raised fabric there is always the possibility, which needs to be checked by laboratory tests, that the adhesion between coating and fabric may be low, leading to various failures in wear. Failure of the sole bond, delamination of the coating starting at the feather edge, and delamination at the inside counter and joint where the two shoes rub against each other in walking, can all arise in this way.

The adhesion of coating test in BS 3424 for coated fabrics is not applicable to polyurethane coatings so that SATRA uses a method which is in effect a sole adhesion test and is similar in principle to the adhesion of finish test for leather (SLTC Method SLF 11). The polyurethane surface of the coated fabric is stuck to a strong flexible material, such as a resin rubber, with a polyurethane adhesive. The two are then pulled apart. The polyurethane coating breaks first and with further peeling the coating separates from the raised fabric at a lower load and remains adhering to the rubber.

RESISTANCE TO FLEX CRACKING

Polyurethane has a good resistance to flex cracking and in the routine laboratory flexing tests for upper materials, such as the SATRA test (PM25), it is usual for the commercial PUCFs to give entirely satisfactory results. The coatings become rather more cracky in the low temperature tests but this is usual for most synthetic materials and there is certainly no reason to regard PUCFs as specially susceptible to cold conditions. Cracking of PU-coated fabrics can occur in wear, however, even when the results of the standard laboratory tests are satisfactory. The types of damage are usually somewhat unusual and detailed studies by SATRA have shown that the causes are defects in the film created during manufacture.

RESISTANCE TO PERSPIRATION

Polyurethane polymers are susceptible to hydrolysis damage, leading to breakdown of the polymer chain and a deterioration in properties. The rate of hydrolysis is increased considerably by the presence of certain substances which act as catalysts, and one of these is urea which is a component of perspiration. Consequently, polyurethane materials in footwear stand a much greater risk of being damaged by perspiration than they would be by pure water.

The hydrolysis susceptibility of poromeric upper material can be checked fairly reliably by SATRA method CM28 which involves flexing the test-piece after the laboratory hydrolysis treatment. With PUCFs, however, we have found this method to be less successful because of differences in hydrolysis resistance between the PU top coat and the PU base coat (or tie coat). It has proved better to assess these two layers separately: the top coat by splitting it off and measuring the reduction in tensile strength due to hydrolysis, and the tie coat by measuring the reduction in coating adhesion.

Our experience at SATRA is that susceptibility to hydrolysis is not a serious problem with PU-coated fabrics in practice. The hydrolysis resistance of the top coat is usually good, and some types of wear damage which formerly we suspected were due to hydrolysis we now find are due to other defects in the coating. The tie coat is more susceptible to hydrolysis but when this occurs, it simply reduces the adhesion of the top coat and does not produce any new types of wear failure. Consequently, when it is evident that the coating adhesion on a worn shoe is low it is desirable to establish whether it was also low in the new material or has been lowered by hydrolysis.

COMFORT

SATRA work has shown that the water vapour permeability of PU-coated fabrics is of about the same order as that of poromerics. Most women do not seem to need permeable uppers to provide comfort in their footwear, but with men it is undoubtedly more important. Consequently, PUCFs are likely to be more acceptable in this respect for men's footwear than the impermeable PVC-coated fabrics.

The figures of permeability we have are for women's weight materials, and if, as we expect, the coatings on men's weight materials are thicker, then it is probably that their permeability will be reduced correspondingly. Also if these materials are backed for reinforcement then there is the possibility that the adhesive coating will also reduce permeability. It might therefore prove that PUCFs, even though better than impermeable materials, are not sufficiently comfortable for a large proportion of male wearers.

H. H. NEUMAIER

Farbenfabriken Bayer GmbH

AQUEOUS DISPERSIONS OF POLYURETHANE IONOMERS FOR COATING AND LAMINATING*

(Received July 31, 1973)

INTRODUCTION

EVERY YEAR MILLIONS of yards of textiles are coated with polyurethane products. Figure 1 shows the percentage increase of textiles coated with polymers (polyurethane and polyvinyl chloride) in Germany during 1968 to 1970.

Let us now discuss the advantages of polyurethane products over PVC coating. Polyurethanes will form elastic films, without requiring a plasticizer, which are fast to dry cleaning and which have very good cold flexibility. Coatings with a solid add-on of approx. 2–3 ounces per square meter are fast to abrasion, breathable and yet waterproof (Table 1).

The di-isocyanate polyaddition process was invented by Otto Bayer, who celebrated his 70th birthday on 2nd November, 1972 [1, 2]. These high molecular polyurethane systems have become so firmly established in textile coating and leather finishing that it is now impossible to do without them.

There are three classes of products on the market: the two-pack polyurethanes, the one-pack polyurethanes and the aqueous dispersions.

Table 2 shows these classes of products in a genealogical tree diagram. These products lead in the textile coating field.

Table 1. Properties of PU Coating Material

1. Flexibility (cold, heat)
2. Abrasion Resistance
3. Tensile Strength/Elongation
4. Elasticity
5. Fastness to
   - Dry Cleaning Solvents
   - Gasoline
   - Oil
6. Adhesion
7. Sorption Behaviour
8. Water Vapor Permeability
9. Resistance to Aging
   - by Light
   - Heat
   - Oxygen
   - Hydrolyses
10. Specific weight

*Permission to publish this paper was granted by the American Association of Textile Chemists and Colorists, Research Triangle Park, North Carolina 27709. The article was presented at AATCC's Symposium on Coated Fabrics, Marriott Motor Inn, Newton, Mass., March 28–29, 1973.

Prod Figures in Mio m²	1968	1969	1970
with PU	9 3 (≈8,5%)	20 (≈14,5%)	38 (≈19%)
Total	110	138	200

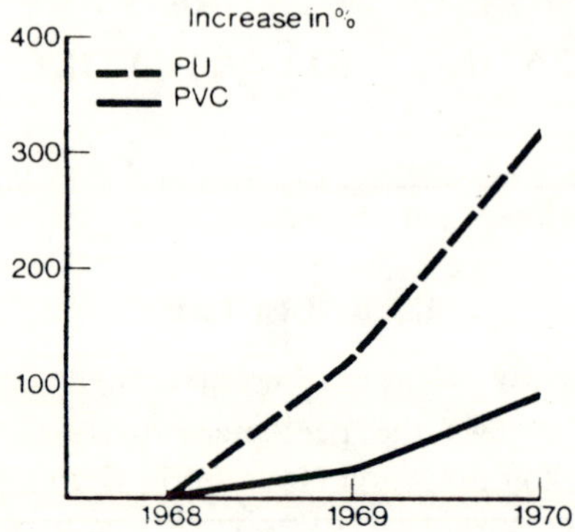

Literature: Kunststoffberater 7,677 (1970) and
Priv. Com. H. Bergs, Farbenfabriken Bayer AG
Leverkusen

Figure 1. Total production of coated textiles Federal Republic of Germany

Table 2.

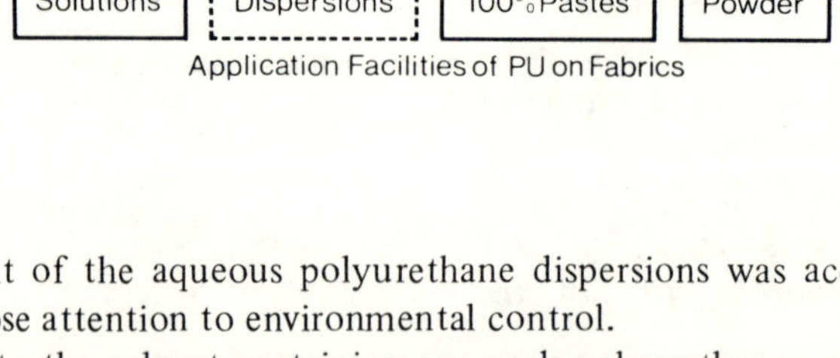

Application Facilities of PU on Fabrics

General

Development of the aqueous polyurethane dispersions was accelerated by the need to pay close attention to environmental control.

In contrast to the solvent-containing one-pack polyurethanes, which will not be discussed in this paper, the pure aqueous polyurethane dispersions contain water as the liquid phase. This does away with recovery of the solvent and with the need for explosion-proof equipment.

The increased energy requirements to evaporate the water of these coatings are partly compensated by the increased solid content of the dispersion. Dispersions can be obtained with 40–55% solid which have good stability. Our developments of such dispersions don't require any emulsifier in order to achieve such high stability.

The high molecular polyurethane dispersions contain hydrophilic ionic centres and hydrophobic polyester or polyether segments. In the aqueous sol condition, we have spherical polyurethane particles which are enveloped by water and which have a particle size of approx. 0.2 microns. Figure 2 shows the uniform size distribution of the particles.

The dispersions are insensitive to heat, mechanical shear forces and diluents, especially since they do not contain any emulsifiers. As a result, the dispersions can be heated, stirred, or mixed with organic solvents. Films are formed already at room temperature. The hardness can be adjusted between Shore A 40 and glass hardness.

Figure 2. Size distribution of PUR-dispersion.

Table 3 lists the other special properties of these products.

Table 3. Special Properties of Impranil Dispersions

Free of Organic Solvents	Long Pot Life
Free of Emulsifier	Mechanical Foam
Simple Pigmentation	Very soft Film and high Elasticity
Combination with other plastic materials	Thermoplastic
	High light Fastness

Chemistry of the Aqueous Dispersions of Polyurethane Ionomers [3, 4]

Aqueous polyurethane dispersions, too, are one-pack elastomers or thermoplasts. The rigid segments of these macromolecules contain ionogenic groups, so that dispersion is possible without using an emulsifier (Figure 3): Di-isocyanate macromolecules [1] can be built up to anionic polyurethane ionomers [3] by means of a chemical reaction with bi-functional chain extenders [2]. It is possible to replace e.g. the aliphatic diaminosulphonic acid [2] as a chain extender, by diamino-carboxylic acid [4]. This results in a carboxylate polyurethane ionomer [5].

EXPERIMENTAL SECTION

Technology of Application of Aqueous Polyurethane Dispersions

Aqueous polyurethane dispersions are eminently suitable, on account of their elasticity, tenacity, fastness to abrasion and cold flexibility, for coating textiles and paper, for leather finishing, bonding of non-wovens, surface coating of wood, and as PVC and gelatine plasticizers and PVC adhesives. Two polyurethane dispersions deserve special mention from the range of the (R) Impranil brands, which can be used either alone or in combinations: Impranil DLN Dispersion and Impranil DLH Dispersion.

Textiles coated with these products can be used in the manufacture of garments such as overcoats, jackets, hats and fancy goods resembling fashionable leather styles, as well as handbag and luggage materials.

Table 4 lists the typical film properties of the two products above.

Table 4. Product Characteristics of Impranil DLN Dispersion, Impranil DLH Dispersion

Typical Physical Properties	*Impranil DLN Dispersion*	*Impranil DLH Dispersion*
Hardness Shore A	55	93
Tensile Strength (psi)	3550	6000
Elongation %	700	670
Set %	10	70
Abrasion: Frank Hauser (mg)	<5	<5
Light Fastness (Blue Scale)	7–8	7–8
400 h Xeno-Test	Low value	good
Hydrolysis Aging 7 d, 158°F 100% RH	very good	very good
Hydrolysis Aging 14 d, 158°F 100% RII	good	very good
Volume Swell in % at R.T.		
Trichlorethylene	550	360
Perchlorethylene	250	91
Water	15	15
Moisture Vapor Transmission ($g/m^2/24$ h)	180	200–280
Cold Flex (Flexible to −40°F)	o.k.	o.k.

Both dispersions are very well miscible. Figures 4, 5 and 6 are graphical representations where the standard products and their mixtures (stated in percent) are

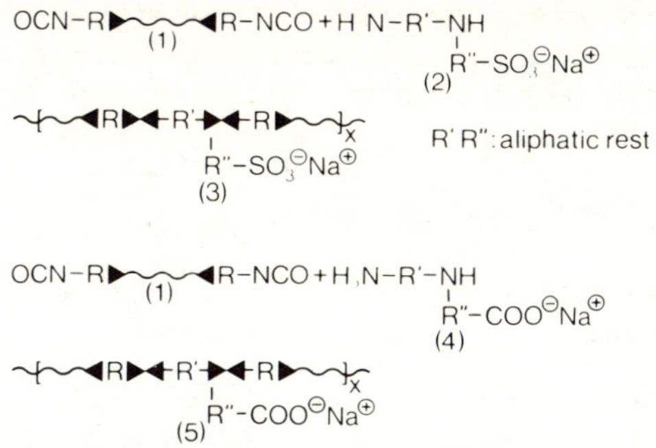

Figure 3. Formation of PU-dispersions.

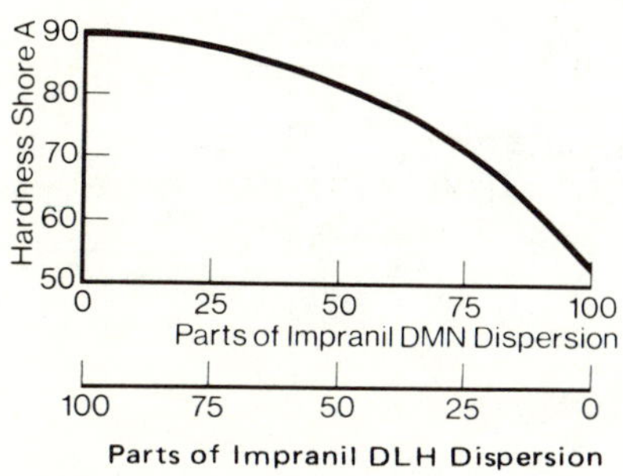

Figure 4.

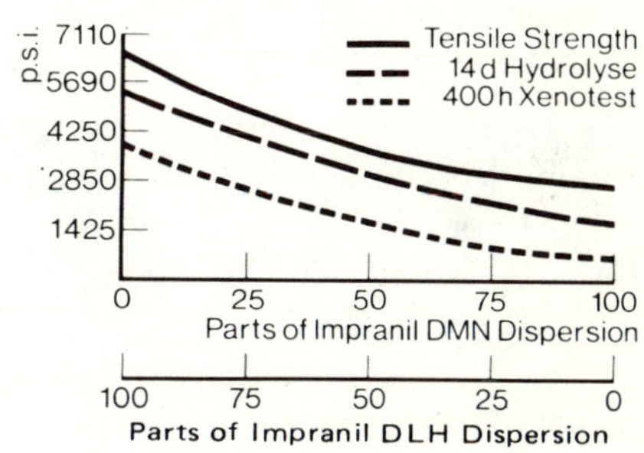

Figure 5.

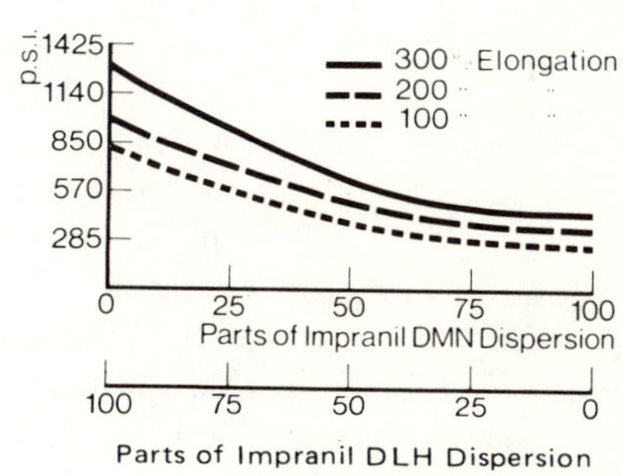

Figure 6.

plotted on the abscissa, while the ordinates show the physical film data according to the composition in each case. With these mixtures, it is possible to adjust the hardness from Shore A 55 to 90. After testing the stability to hydrolysis for 14 days, and after a Xenotest exposure for 400 hours, there will be a change in the stability of the films with increasing Impranil DLH Dispersion content. The same applies likewise to the modulus of extension.

The characteristics of the products are shown in Table 5. Chemically speaking, we have anionic, aliphatic polyurethane dispersions. These are supplied as white, approx. 40% dispersions of low viscosity, and have a storage life of 6 months.

The particle size is approx. 0.2 microns, the specific gravity approx. 1.1, while the pH level is approx. 6 and the viscosity about 220 cps. The product must be

Table 5. Characteristics of Impranil-Dispersions

Form:	Anionic 40% T.S. aqueous dispersion
Chemical Composition:	Aliphatic polyester urethane
Color:	Milky, white
Specific Gravity:	1,1
pH:	Approximately 6
Particle Size:	0,1–0,2 micron
Viscosity:	220 cps (Brookfield 20°C)
Storage Life:	6 months

stored at a temperature above 3°C to prevent freezing.

TECHNICAL ADVICE ON PROCESSING

Thickeners

The 40–50% anionic aqueous polyurethane dispersions, which have a low viscosity, must be thickened to a viscosity of 7000 to 8000 cps. This can be achieved by adding polyacrylate copolymer dispersions such as Mirox AM (Stockhausen), Natrosol types (Hercules Powder), or an aqueous vinyl pyrrolidon solution such as Impranil Thickener PW.

Pigments

The aqueous polyurethane dispersions can be colored by stirring in suitable pigment pastes. This is a straightforward procedure, without requiring any grinding. The following pastes are suitable:

- Impranil PP 800 Colour Pastes
- Levanox Fine Pastes
- Fast Paper Black T
- Acramin Color Pastes
- Baysin DD Colour Pastes

These are inorganic or organic pigments. Furthermore, the Baysin DD Colour Pastes have a very low emulsifier content. Normally, pigmentation up to 10% will be sufficient (in relation to the Impranil dispersion).

Fillers

A dry handle of the top coat produced with Impranil DLN Dispersion, can be obtained by adding fillers, e.g. China clay, Surfex MM (Diamond Chemical Company) or Calibrite (OMYA), in quantities up to 10% filler in relation to the Impranil DLN Dispersion. Fillers are not recommended for use in the adhesive coat, since this will reduce the adhesion.

Mixtures of Impranil DLN Dispersion with polyacrylates of different hardness can be prepared with a maximum acrylate content of 15%, e.g. Acralen AFR,

Acralen AMR, Acralen BS, and Acralen BSH. If the additions are increased above 15%, satisfactory cold flexibility of the coating is no longer guaranteed.

Preparation of the Coating Pastes

Pastes must be prepared with great care. Figure 7 shows the important steps.

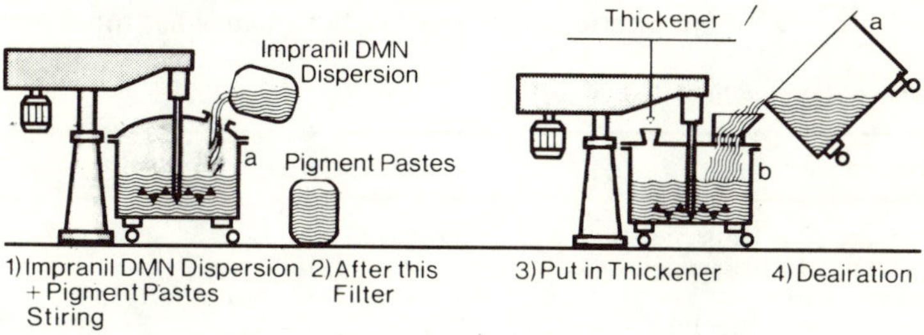

Figure 7. Mixing of Impranil DMN dispersion.

The pigment colour pastes are added to the aqueous dispersion by stirring into the vessel a. To remove all foreign matter, the mixture is filtered into vessel b. The width of the meshes of the filters used in this filtering process, is generally 30–50μ. Thickener is added by means of a vacuum stirrer which is placed on the mixer, and the paste is de-aerated while stirring. This de-aeration is necessary for the top coat. Any presence of air in the paste will cause pinholes in the surface of the coating (Figure 8). De-aeration is not necessary in the case of adhesive coats.

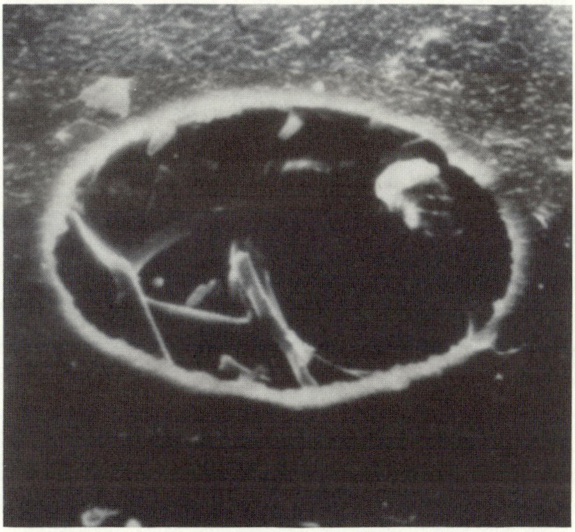

Figure 8.

The ready coating paste should have a viscosity of 7000–8000 cps (measured with a Brookfield LVF viscometer at 20°C, spindle No. 4 and 30 r.p.m.).

Formulations

Table 6 gives the formulations for top and adhesive coats. Coatings executed with Impranil DLN Dispersion or with Impranil DLH Dispersion must be finished, because they are of a thermoplastic nature and exhibit a somewhat rubbery surface.

Table 6. Formulations

Product	*Skin Coat*		*Adhesive Coat*
Impranil DLN Dispersion	1000 g		1000 g
Impranil DLH Dispersion		1000 g	
Pigment Pastes	50–100 g	50–100 g	0–50 g
Impranil Thickener PW	20–30 g	20–30 g	20–30 g

Here is an example of a finish on aqueous dispersion basis: 1000 g Bayminol-Binder AF, 0–200 g Pigment Pastes and 20 g Impranil Thickener PW.

It is also possible to coat with finish products which contain solvents, e.g. Impranil D-628, Impranil D-628 Matt, or mixtures thereof.

Coating Technique

Coating with aqueous polyurethane dispersions is carried out with the equipment currently used in textile coating. This equipment need not be explosion-proof. A Transfer Coating Line with two Coating Heads is recommended. Release papers should have good wet strength properties. Such papers are, for example:

Stripcote AR C1S (manufactured by S. D. Warren)
Feikostrip BNW (manufactured by Feikes, Germany)
KTFS-10 (manufactured by Riegel)

The coating paste can be applied by means of a knife-over-roll or a reverse roll coater. The use of a knife-over-rubber blanket will not guarantee uniform coating.

When using a coating knife, care should be taken that the distance between the knife and the release paper does not exceed 100–150μ in the top coat, since cracks can occur in the film during drying if this film thickness is exceeded.

If the adhesive coat is applied using an aqueous polyurethane dispersion, the gap of the knife must be below 150μ. The required handle can be obtained by varying the laminating nip. Penetration as well as strike through of the coating paste should be avoided because this will harden the handle.

Figure 9 shows a Transfer Coating Line with two Coating Heads:

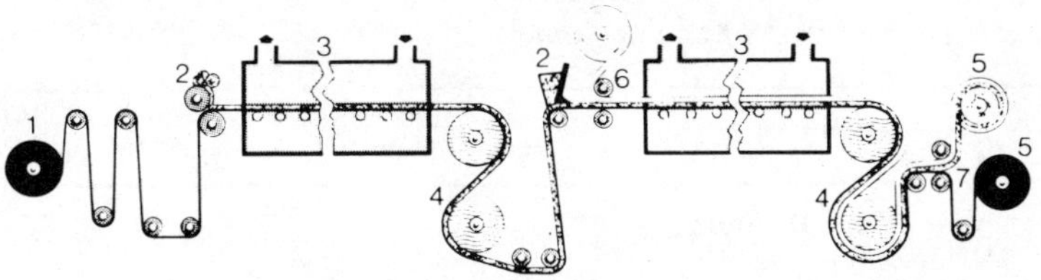

1. Take off, 2. Coating. 3. Drying. 4. Cooling Cans. 5. Batching. 6. Laminating Rolls. 7. Separation

Figure 9.

The ready paste is applied to the release paper, for instance, by means of a reverse roll coater. The next stage in this process is the drying channel in which the temperature is made to rise from 200°F via 230°F to 265°–290°F. The top coat runs over a cooling can and further guide rolls to a second coating head. The adhesive coat is applied at the second coating head; the substrate is laminated at this stage, and drying is effected already at 250°F. The temperature can be raised up to 290°F.

The coated material is separated from the release paper and then finished with one of the finishing agents mentioned above. This can be done by either printing or the use of a doctor blade. The polyurethane dispersions can also be applied by direct coating. A thin coat for the first coat is recommended; so that the dispersion will not penetrate into the base.

Heat-sealing Properties

The polyurethane film is plasticized at a temperature of 346°F. Due to this thermoplastic behaviour, coatings with Impranil Dispersions can be heat-sealed.

Embossing

The coating can be embossed on a normal embossing calender. In this case, the coated side is heated to the plasticizing temperature in an infra-red pre-heating zone, and embossing is carried out with a cooled roll.

Properties of Substrates Coated with Polyurethane Dispersions

Examination of the coatings with our polyurethane dispersions have given outstanding results of a quality which has so far appeared unattainable for aqueous polyurethane systems.

Table 7 shows the results of tests carried out in actual mill runs.

After the goods were washed in the washing machine at 140°F, the coatings showed slight swelling, but the surface was still fully elastic and the adhesion in the wet state was satisfactory. A slightly swollen surface is also observed after dry cleaning.

Table 7. Test Results of Coated Fabrics

Tests	Skin Coat DLN Adhesive Coat DLN	Skin Coat DLH Adhesive Coat DLN
Abrasion (Frank Hauser)		
(DIN 53528) mg weight/loss	<5	<5
Cold Flex (DIN 53513)	−40°F	−40°F
Cold crack at −40°F (DIN 53361)	o.k.	o.k.
Endurance stress (DIN 53359)	>100 000	>100 000
Ballyflexometer at RT	>100 000	>100 000
Ballyflexometer at 14°F	>30 000	>30 000
Moisture Vapor Transmission ($g/m^2/24$ h)	180	200–280
Adhesive kp/cm		
a) dry	1.6–1.8	1.6–1.8
b) wet	1.1–1.2	1.1–1.2

Mechanical Foaming of Aqueous Polyurethane Dispersions

As already stated, the aqueous polyurethane dispersions can be foamed mechanically. This opens up new possibilities of application without the use of emulsifiers.

A polyurethane disperse foam has two simultaneous functions:

1. The foam replaces the usual pile volume of a substrate which has been raised with great expenditure.
2. At the same time, the foam serves as adhesive coat for the compact one-pack top coat.

If a low-priced base is used, this will give a very interesting handle of the kind which is at present very popular in the apparel field. Foaming can be done with e.g. 1000 g Impranil DLN Dispersion and 30 g Impranil Thickener PW.

In laboratory trials, foaming can be done by means of a kitchen mixer. Foam mixers for bulk work are supplied by e.g. Eur-o-matic (Weesp, Netherlands) and Oakes (Macclesfield, England). A foam mixer of this type is shown in Figure 10.

The essential conditions for a good-quality foam are a very fine pore size, and a suitable weight per litre. The weight per litre can be controlled by means of the pressure in the system and by varying the air supply. The lower the pressure in the system, the higher the litre weight of the foam. The pressure in the system depends, among other things, on the length of the tubing, the tube diameter, and the air supply. A very suitable pressure is approx. 2.2 atmospheres pressure. This will produce the required litre weight of 500 g with an air supply of 0.5 scale divisions. If the foams are too light in weight, the amount of solid added to the base will be too low. This will result in poor adhesivity.

Figure 11 shows how the foam is applied before the coating knife. The coloured top coat produced with Impranil D-628/D-398 has already been applied. It is recommended to use a release paper with a fine grain.

As in compact coating, a roll coater can be used. 400–500mμ is the optimum

Figure 10.

amount of foam applied. This corresponds to 2–3 oz solid per square metre. The surface of the ready coating is finely grained if these add-ons are used. The substrate – which should have the lowest possible shrinkage with a weight of 150 to 200 g/m^2 – is then laminated to the coated material (Figure 12).

The distance between the laminating rollers depends on the amount of foam and on the thickness of the substrate. The sum of both plus 0.1 mm gives the distance between the laminating rollers which guarantees a good quality of the goods with good adhesivity.

Drying is initially effected at 194°F. In the second zone, the temperature is raised to 250°F. The temperature in the final zone should be approx. 320°F. Good curing is necessary to ensure good adhesion to the top coat and to the base. The coated material is separated from the release paper. Final finishing treatment can be done in several ways. The disperse foam can also be applied by the direct coating process, in the same manner as in compact coating.

Lamination

The aqueous polyurethane dispersions are suitable for producing laminates

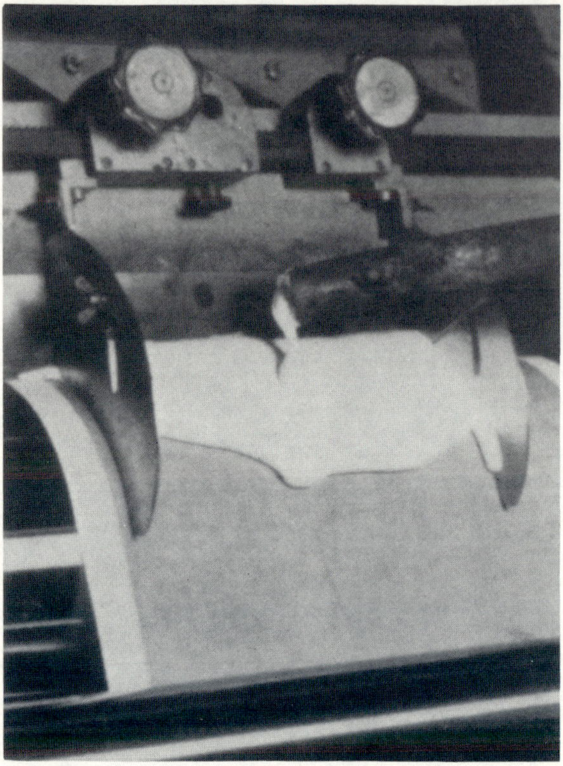

Figure 11.

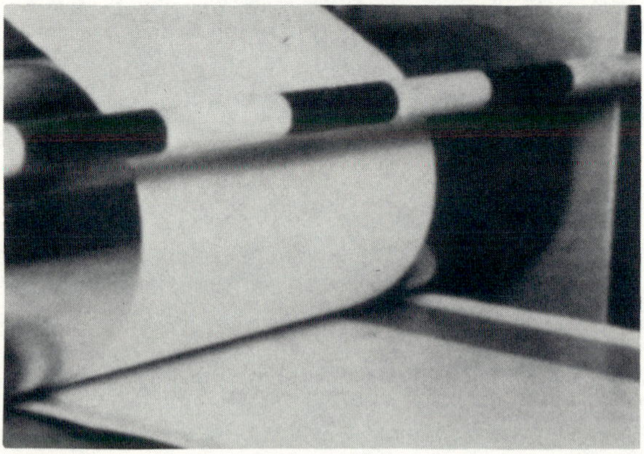

Figure 12.

which are fast to washing and to dry cleaning, both in textile-to-textile bonding and textile foambacking. Technically, this is done in a simple manner by applying the thickened dispersion to the textile or the foam by means of a roller. The foam is bonded to the textile material by light pressure.

A suitable formulation is e.g. 1000 g parts Impranil Dispersion and approx. 30 g parts Impranil Thickener PW.

CONCLUSION

This paper has attempted to describe the present state of the application of aqueous polyurethane dispersions for textile coating and laminating.

When these entirely novel products were developed, particular attention was paid to the requirements of environmental protection.

Top-quality products of the aqueous polyurethane dispersion type which will remain stable for a very long time, without any emulsifier additions, are now being marketed. They can be used both for top and adhesive coats in compact coating, and for adhesive foam coats. The transfer coating process is the method of choice.

Bayer has been successful in developing aqueous polyurethane dispersion products for textile coating and lamination which are of the same quality as the solvent-containing polyurethane systems.

With these polyurethane dispersions, new avenues for the market are now opening up which will certainly be rewarding.

REFERENCES

1. DRP (German Patent) 728 981, I.G. Farben Ind. O. Bayer et al.
2. O. Bayer, Angew. Chem. *59* (1947), 9, 257–272.
3. D. Dieterich et al, J. Oil Col. Chem. Assoc. *53* (1970), 363–379.
4. D. Dieterich et al, Angew. Chem. Int. Ed. *9*, 40–50 (1970).

DAN PSEJA

Millmaster Onyx Corporation
Urethanes Specialties Division
624 Schuyler Avenue
Lyndhurst, New Jersey 07071

NEW URETHANE PRODUCTS FOR THE TEXTILE INDUSTRY*

(Received January 24, 1974)

KEY WORDS: Urethanes, coatings, shoe uppers, urethane latex, adhesives.

INTRODUCTION

THE USE OF urethane coatings has been experiencing a tremendous growth in the past few years. Urethane coatings made their largest penetration into textile coating in the mid to late 60's with the popular "wet look" fashions. Since then, transfer coaters have found the characteristics of urethanes an asset in entering the more performance-oriented markets of upholstery and shoe uppers. With a diminishing supply of leather, and increased leather costs, the urethane shoe upper market has been growing rapidly. Reports that 50% of all shoe uppers will be manmade by 1975 have been published. Urethanes will make up a large portion of that 50%.

Urethane coatings have not only found increased markets in transfer coating, but have made large strides in other textile-oriented industries. Their outstanding abrasion resistance and flexibility have made them popular as topcoatings, particularly in the topcoating of vinyls. Urethanes have, in many cases, replaced vinyl and vinyl-acrylic topcoats.

The automotive upholstery industry is making efforts to develop softer, more leather-like vinyl upholstery. This, of course, means higher plasticized vinyls and need for improved adhesion and abrasion resistance. Low temperature flexibility and resistance to plasticizer migration are other requirements. Urethane topcoats have enabled the automotive industry to obtain these results while also obtaining benefits of stylish and aesthetic efforts due to the compounding versatility of these solution coatings.

Another market that opened up the need for superior topcoats was developed by

*Presented at the Urethanes in Textiles Seminar, Charlotte, NC, January 24, 1974, sponsored by Technomic Publishing Co., Inc., Westport, CT.

the transfer coating of urethanes for shoe uppers. In order to obtain the needed scuff resistance while maintaining softness and flexibility of the shoe upper composite, a thin, highly abrasion-resistant urethane topcoat was needed. Conventional topcoating either did not have the abrasion resistance or lacked the flexibility.

NEW URETHANE COATING

The Urethane Specialties Division of Millmaster Onyx Corp. has designed a specific polymer to provide adhesion to vinyl substrates without the use of high boiling solvents or primer coating. Our Urethane Topcoat LS 370 has been found to provide good abrasion resistance and dry surface properties without the loss of flexibility needed to both the shoe upper and upholstery marketplace. Oftentimes the amount of material necessary to achieve both these performance characteristics with conventional topcoating materials, necessitates more than one pass through the equipment. Therefore, Urethane Topcoat LS 370 has been synthesized at a higher solids than conventional topcoating materials. The uncut material at 30% total solids gives more flexibility to the coater in order to utilize standard gravure coating equipment and still obtain higher deposition of material on one pass. Higher patent finishes are also possible due to the increased amount of a clear topcoat of LS 370 attainable in one pass. An additional benefit is the lesser amounts of solvent emitted into the atmosphere.

The urethane topcoat is capable of a greater degree of compounding while still maintaining other physical properties. For example, it can be flatted to a much greater degree than vinyl-acrylic topcoatings and still is capable of passing stringent abrasion-resistance testing. A flatting concentrate is offered that can be added to our polymer to give the compounders more latitude in obtaining various lustres.

FLAMMABILITY CONSIDERATIONS

With the increased degree of emphasis on safety, there is need in the textile coating industry in general, to develop coated substrates that are fire retardant. Previously, many coatings have contributed to the flammability of their substrates. The emphasis of developing composites that will be fire retardant is a problem for the manufacturers of urethane coatings and their users. Problems such as solution stability and maintenance of physical properties have heretofore been obstacles in developing such coatings. The industrial direct coatings market has been looking for good fire-retardant urethane coatings particularly for the direct coating of nylon. End-use applications such as airline escape chutes, outdoor tenting and waterproof outerwear are the intended markets. In all cases, it must be the responsibility of the coater to determine and maintain the fire-retardant properties of the resultant coated goods. The importance of testing cannot be overemphasized.

Efforts to approach this need have resulted in the Urethane FR 130 series of polymers. These polymers are specifically synthesized to minimize contributions of flammability to their coated substrates. They are available in different viscosities to accommodate particular coating requirements. They are also available in several hardnesses to vary the hand of the coated goods. It has been found that the FR 130 series of polymers will not break down when subjected to extended ultraviolet exposure.

In one particular case, for example, a 1.7 ounce nylon coated on both sides with a total of 5/10 of an ounce of Urethane FR 135 has passed the CPAI-84 vertical flame test. We noticed about a 3½- to 4-inch char with no flame present after the 12-second burn.

Utilization of Urethane FR 132 and 133 polymers in airline escape chutes does not affect the conventional heat sealing or air holding properties needed for their application.

100% SOLIDS URETHANE SYSTEMS

100% total solids coatable urethane systems are being developed and commercialized. Previously 100% total solids urethanes have been utilized in the castable, foam and powder coating industries. With more and more stringent environmental requirements being placed upon coaters of solvent-based polymers, the 100% solids concept is quite advantageous. Recently the shortages affecting solvents utilized in urethanes has also added a push to develop such coatings. The coater can convert all of the purchased compound to coated goods. 100% solids urethane coatings systems also have other advantages. Thicker coatings may be applied in one pass where high coating weight is required. This is important for transfer coaters selling to the upholstery and shoe upper markets as well as for the industrial direct coatings market. Also since no solvent must be removed from the coatings, difficulties arising from solvent boiling and entrapment are virtually eliminated.

While there are significant benefits, there are some aspects of 100% solids coatings which are disadvantages. The mere nature of utilizing a chemically reactive system and attempting to deposit films with thicknesses that must have close tolerances would lead one to visualize some of the problems that could be encountered with such systems. First, technical supervision is often needed to a much greater degree in operating 100% solids coatings lines. Equipment utilized for proportionating, mixing and distribution of such systems must be designed with extremely close tolerances. Our division has been exploring 100% solids urethane coatings systems for the past few years. We have designed a transfer coating system for upholstery applications. Not only did we have to design the polymer system, but pigmentation and preparation of suitable vehicles to carry such pigments had to be studied. Through our experience of polymer design, we also found it necessary to design and construct proportionating, metering and mixing machinery. This was necessary in order to develop the greatest inter-relationship between polymer and equipment. We learned through our experience that metering and mixing was extremely important in having a successfully coated product.

We have taken our pilot proportionating equipment and have run continuous coating operations on a full-size production line yielding coated goods of up to 30 inches in width. In order to coat full widths our pilot proportionating would have to be modified for increased output.

The author's company is devoting research and development to eliminate some of the intricacies of the 100% solids urethane coating systems as they now exist and are also exploring UV curable systems.

URETHANE LATEX

Another new urethane product that has been designed to meet current needs in industrial applications is that of a urethane latex. Basically, the urethane latex is a water-based emulsion with some sort of emulsifying system. Some latices are designed with additive emulsifiers while others are self-emulsified. Their particle sizes range from about 0.1 microns to several microns in diameter.

PRODUCT DESIGNATION IS URETHANE LATEX SHS

Our history with urethane latex started early in our growth in the urethane industry. We have found that this was perhaps premature for the industry at that time since most coaters that were using solution urethanes were not accustomed to handling a water-based system. In some cases, heating requirements had to be modified in order to completely remove the water. Other difficulties existed in the handling and compounding of urethane latices. Coaters also found a difference in rheological properties as compared to the solvent-based urethane coatings.

Water-based urethane latex is a natural for this day and age of shortages of solvents and environmental restrictions. Since no solvent is used, the price per dry pound of polymer is substantially lower than conventional solution coatings. In transfer coating applications, Urethane Latex 5HS has been utilized as an adhesive between the skin and the substrate. It has been found that problems of solvent attack on the urethane skin are virtually eliminated. The performance of urethanes carried on to their latex counterparts by imparting low temperature flexibility along with high tensile strength, uncommon to many other polymer latices. Urethane Latex 5HS is a cationic emulsion and has a good affinity to cellulosics. This generally leads to higher bond strengths than obtained with solvent-based urethane adhesives.

Our urethane latex has found a home where unusual bonding properties for hard-to-adhere substrates are encountered. Plasticized vinyls have generally presented a problem in attaining good bonds. We have found it has good resistance towards plasticizers in a variety of vinyl compositions utilizing the latex as an adhesive. Other adhesive areas where the urethane latex has found application are for the bonding of nylon, polyester, fiber glass, aluminum, Mylar and both urethane and vinyl foams. Generally, we find that less urethane is needed to outperform cheaper competitive adhesives. Therefore, utilizing urethanes can be quite cost competitive with less expensive types of adhesives. This has opened the door for applications as adhesives and beater additives in the paper industry as well as saturants for nonwoven substrates. Generally we have found that a 10 to 15% add-on during saturation of nonwovens improves dimensional stability and strength to the extent that this substrate can be utilized for applications such as backing for shoe uppers.

In one paper application the Urethane Latex 5HS has been found to perform well in the manufacture of wallpaper where the latex is applied to a wet stock and bonded to a backing of either cotton scrim or synthetics such as polyester nonwovens. Less material is needed to get a better bond and obtain greater dimensional stability.

Experience with Urethane Latex 5HS in the transfer coating industry has shown it to be an excellent dry cleanable adhesive for apparel applications where a soft "Cabretta" feel is desired. In shoe upper applications we have also found that the urethane latex will yield bond strengths of 10 to 18 pounds per linear inch versus 6 to 10 pounds per linear inch utilizing good solvent adhesives. This improved bond strength to napped cotton, for example, tends to give much improved abrasion resistance. In a composite using the same thickness of skin and a comparably good urethane solution adhesive, Tabor abrasion with an H-22 wheel and 1,000 gram weight to be 2,000 cycles compared with 200 cycles for the solution adhesive.

Hydrolytic stability, of course, is important for the shoe upper industry. Composites subjected to 2 weeks of aging at 95% relative humidity and 158°F show no change in Tabor abrasion and only slight loss in Bally flex results. Our bond strengths decrease only about 25%.

Other adhesive applications with urethane latex include the laminating of calendered vinyl to urethane or vinyl foam. Composites of calendered urethane with textile substrates have also been laminated in equipment that could not normally handle solvent adhesives due to either poor ventilation or nonexplosion proof design.

The author's company is developing a total transfer coated urethane latex skin and adhesive. We have achieved such a system presently utilizing a solution urethane lacquer topcoat or a silicone surface treatment, in order to impart wash resistance. All urethane latices have a problem of water spotting to some degree due to their emulsifying system. Better water resistance is an avenue that is being explored to improve washability of urethane latices.

Manufacturers of urethane coatings have been looking to the needs of industry in developing new products to satisfy new applications. As evidenced by the growth of urethane manufacturers over the past few years, industrial applications are utilizing some of these new developments. With continued interaction of polymer manufacturer and end user, the outlook is bright for urethane products to fill the needs of higher performance requirements in the textile industry.

DR. K. H. STEIGERWALD

Steigerwald Strahltechnik
Munich, Germany

THE INTRODUCTION OF PERMEABILITY INTO PVC AND POLYURETHANE UPPER MATERIALS BY ELECTRON BEAM PERFORATION

(Received January 2, 1974)

ABSTRACT: A new principle for producing poromerics is proposed in which an electron beam is used to perforate composite materials with an outer layer of PVC or PU. The advantages lie in the fact that the perforated material allows air from inside the shoe to be more readily exhausted to atmosphere, carrying moisture with it and enhancing foot comfort. Tests so far indicate a comfort level at least as good as natural leather. The concept of a production machine is discussed and properties of materials and economics are presented.

Key Words: Poromerics, permeability, shoe uppers, electron beam perforation.

INTRODUCTION

ELECTRON BEAMS ARE characterized by their extreme speed, precision and controllability and these properties are being used by Steigerwald Strahltechnik to provide new manufacturing capabilities in many industries. Among our development programmes is a large effort aimed at bringing the properties of the electron beam to bear on the problems of poromerics and our research work has now brought us to the point where all the difficulties can be defined and the solutions to them can either be proposed from todays' knowhow or extrapolated with a high degree of confidence. Nevertheless, about two more years' development and manufacturing time is needed before full-scale production machines can be placed in shoe-producing factories.

PRINCIPLES OF GENERATING ELECTRON BEAMS

Various materials, if heated to a sufficient temperature, emit free electrons which can be collected, accelerated and focused by electrical means into an immensely powerful and very fine beam capable of melting and penetrating most materials known to man. Powerful here means in the sense of energy density rather than total power. A typical electron beam generator today consumes about 25 kW and produces a beam of 1 to 15 kW continuous rating capable of being focused into a spot of, say, 0.1 to 0.5 mm diameter. Consequently the energy density in the spot lies in the range 10^6 watts/cm^2 to 10^8 watts/cm^2.

FEATURES OF A PRODUCTION MACHINE

There are two ways in which machines could be used to process material; either by perforating a continuously moving band from which uppers could later be cut or by perforating uppers after they have been fashioned.

One of the most interesting possibilities is to combine the electron beam process with the production of flow-moulded uppers; the advantage of producing flow-moulded shoes is that complete and decorated uppers or parts of uppers can be produced from a single piece of material. The process is applicable to PVC-coated fabrics, some PU-coated fabrics, some leathers and some poromerics, achieving a very high fidelity reproduction of an original leather or other surface. The effects which can be reproduced include design features such as overlays, stitching, cut-outs, perforations and weaving, as well as the original surface texture.

Now if the uppers can be processed by electron beams to give good comfort properties, then the two new processes in combination could radically affect the whole concept of shoe manufacture during the coming decade. It has been our experience in other industries that the application of electron beams may cause completely new thinking on production policy. In the hack-saw industry, for example, a long-standing policy of labour-intensive production techniques was changed to a capital-intensive programme in the remarkably short space of two years by the European market leaders. The benefits were a better product at lower overall manufacturing costs but the impact on capital investment planning was severe. If electron beam perforation proves, as we predict, to be equally important in the shoe industry, then that industry will also be obliged to re-appraise its thinking about capital investment in plant.

The perforation process with electron beams is effected by local heating, melting and vaporizing of the material. For 'artificial leather' by which is meant those man-made materials which are designed to have leather-like properties, it is imperative that very fine perforation speed is required to produce economically perforated 'artificial leather.' In spite of the fact that the electron beam produces perforations at very high frequencies, in the case of 'artificial leather' we have to work with a multi-beam generator which allows many fine-focused electron beams to be directed on the material at the same time.

Figure 1 shows a scheme of the perforation process. A great number of electron beams which in the scheme are illustrated as focused cones are directed on the 'artificial leather' in a straight line similar to a comb. The upper to be perforated is supported by a plate which is moving relative to the stationary electron beam. The electron beam is pulsed and produces holes in straight lines.

Figure 2 shows a diagrammatic scheme of an electron beam machine which gives an idea of the operating concept for the perforation of flow-moulded uppers. The

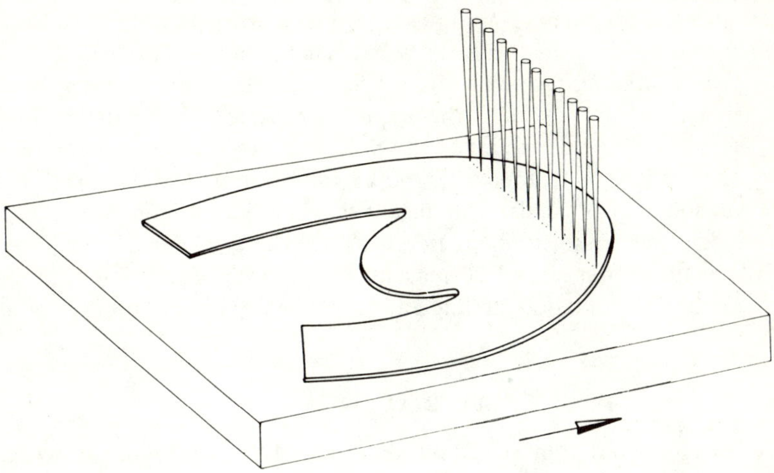

Figure 1.

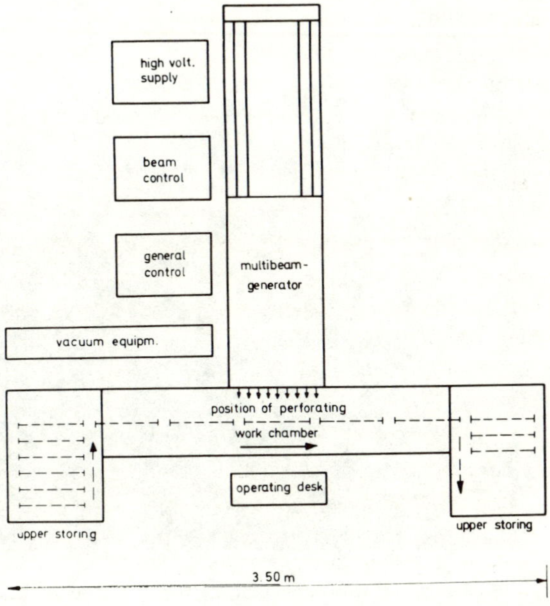

Figure 2.

multi-beam generator, the work chamber as well as the upper storing chambers present together one vacuum chamber. The upper storing chambers are connected with the work chamber by valves which cannot be seen here. For the process, uppers are inserted in containers into the left upper storing chamber. This has to be evacuated and to be connected with the work chamber by opening the valves. Then the uppers are continuously moved through the work chamber into the right upper storing chamber and on the way the uppers are perforated in the perforation station. The machine is working continuously and no interruption of the work occurs by the feeding with uppers. The floor-to-floor-time for the perforation of one upper is about 10 seconds. This means that the output amounts to 550,000 pairs of uppers per year. In this case the machine is intended to work on a 2-shift-basis at utilization rate of 70 percent. This type of machine is expected to cost in the order of £200,000 sterling and to perforate uppers at a cost of about 15p each.

MATERIAL DATA

Most of our work has been carried out on PVC and PU outer skins, attached to a variety of linings.

Figure 3 shows a micrograph taken by a scanning electron microscope of a perforation in PVC coated on a non-woven fabric lining.

We expect to provide a material which after perforation, allows the movement of the foot in the shoe to give a pumping action, air being continuously exchanged through perforations and exhausting moisture-loaded air from inside the shoe to atmosphere.

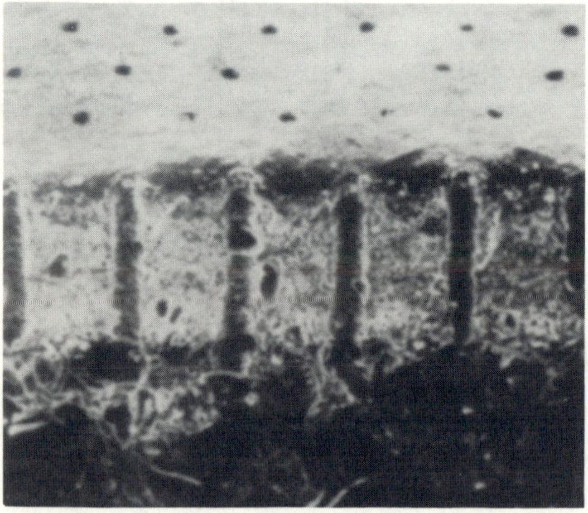

Figure 3. EB-perforation of synthetic leather (PVC).

thickness	*0.8 mm/1.3. mm*
hole diameter	*0.04 mm*
hole density	*2000 L/cm²*

From various tests which we have undertaken the following data may be of interest:

hole diameter:	in the range 0.05 to 0.2 mm
hole depth:	up to 1.5 mm
hole density:	50 to 500 holes per square centimetre
Gas Permeability at pressure difference of 20 mm of water	in the range 10–50 $cm^3/s \times 10$ cm
Water Vapour Permeability (after Mitton)	in the range 2–10 $mg/h \times cm^2$

Selection within this range can be controlled by the combination of hole diameter and density. It may well be possible therefore, to design shoes for specific climatic conditions such as European winter, European summer, tropical summer etc., by varying this parameter.

Water Permeability

Holes of the size – 0.05 to 0.2 mm – tend to repel water owing to surface tension effects. Treatment with a hydrophobic material will enhance the water-repellent characteristics without noticeably affecting "breathability" and tests at a pressure of 40 mm of water indicate that more than 30 minutes will elapse before penetration of water to the inside of the material is observed.

Flexometer and Tensile Tests

In general it may be stated that the loss of mechanical properties will be less than 5% of the unperforated material strength. This low loss value is due to the fact that only the thin PVC or PU skin is perforated and the lining material remains whole.

Wear Trials

The results so far are very encouraging and indicate a level of comfort at least as good as that provided by leather. An improvement to a level superior to that of leather may be achieved if the pumping action referred to earlier is borne out, on a subjective basis, by exhaustive wear trials. Figure 4 shows a pair of men's flow-moulded PVC-upper shoes.

CONCLUSION

Several years' research work in our own laboratories have convinced us that a new approach to poromerics can be gained by employing an electron beam to perforate existing PVC and PU materials. Discussions with various objective

Figure 4. Men's shoes made from flow-moulded PVC-uppers.

authorities such as SATRA have confirmed the value to the footwear industry of such new materials and we have recently entered an arrangement under which the first machines will be produced for and tested by Salamander. This will take the process through to normal production of perforated uppers at an economic price. This phase of work is expected to last about two years.

D. J. MORLEY*

FIRA Research Department

POLYURETHANE COATED UPHOLSTERY FABRICS – FAULTS THAT CAN OCCUR

(Received January 2, 1974)

ABSTRACT: This article classifies faults that have occurred with polyurethane fabrics in upholstery. Recent correspondence elsewhere has highlighted difficulties that can arise in use but it remains true that, with careful attention to selection of material and source of supply, ensuing quality control procedures applied regularly, and making up techniques, PUs will continue to hold an important place within furniture fabrics. However, care must be exercised in choosing and inspecting a fabric.

INTRODUCTION

POLYURETHANE- (PU) COATED upholstery fabrics are basically fashion materials in which the leatherlike feel and soft handle are obtained by bonding a thin polyurethane coating to a raised cotton backing fabric. Although some materials have given consistently good results, the production of consistent materials from batch to batch of a given quality is not easily achieved and faults have occasionally occurred to a degree which has been commercially unacceptable. It is our experience that faults have been due to a variety of causes, some of which can be deduced from a superficial examination of returned covers; some others are not as easily assessed and may have more than one cause, particularly where a cracked appearance results on the coated surface of the material.

For many of the superficially similar faults found during our examination of returned covers it is a great advantage to be able to inspect the covers, both on the coated surface and in the thickness direction of the coating (i.e. in cross-section) under a magnification of up to X 100, using a binocular reflecting microscope which possesses objective lenses with a good depth of field. This enables flaws in the coating to be seen in detail which would escape notice under normal inspection.

Most of the faults noted at FIRA have arisen because of minor defects in the coating which occurred during manufacture and a few have occurred wholly or partly because the coatings have aged.

One of the most consistent complaints from upholsterers is that the surface appearance of PU-coated fabrics can vary from batch to batch of a given quality

*Mr. Morley is a principal research officer with the U.K. Furniture Industry Research Association, at Stevenage, England, whose responsibility is within upholstery fabrics – developing test methods, fault finding and material selection.

and colour. This is a valid complaint, but it does not directly affect the performance of the material in use. It arises simply because a thin coating tends to deform to follow the contours of the weave pattern of the base fabric, to a degree that depends on the tackiness of the coating and the lamination pressure used to combine the fabric and coating. It is unlikely to occur with the thick coatings used in expanded vinyl-coated fabrics.

SEAM SLIPPAGE

This fault occasionally occurs with the softer polyurethane-coated fabrics especially where seams sewn across the width of the material run across the main direction of the arms or are piped. The seams gape open slightly and the coating can split between stitch holes so that it comes apart along the seam to expose the basecloth. The fault is particularly noticeable if the basecloth is of a different colour to the coated surface, and it occurs primarily because the weft (widthways) yarns of the basecloth are relatively soft yarns and form the raised surface of the fabric. It is this surface which is coated and the type of weave pattern used to produce this effect is somewhat prone to slip; the weft yarns slide along the warp (lengthways) yarns when widthways seams are stressed during upholstering. Seam slippage can be minimised by the use of correct stitching procedures, which are set down in another section of this article.

DELAMINATION

Delamination occurs when a thin layer of PU coating separates from the basecloth and looks rather like a deflated balloon. This fault, which is not as common as it used to be, may be present on soft materials even before they are used. On chairs it appears to occur more frequently adjacent to seams. Materials which possess a soft coating which tends to "float" on the base fabric are particularly prone to delaminating through poor adhesion between the basecloth and coating. Such materials should not be offered or chosen for upholstery, and do not usually give a realistic leather "break" when the surface is deformed. Once the coating has delaminated it is likely to be damaged by scuffing and snagging, which can peel off substantial areas of coating and expose the base fabric.

SURFACE IMPERFECTIONS IN THE COATING

Faults can arise where the surface is contaminated with small holes (pinholes) or because cotton hairs from the raised basecloth protrude through the PU coating.

Faults from pinholes are generally more serious, but by no means all fabrics which show either pinholes or protruding hairs are liable to be faulty. Both types of imperfection are best seen (pinholes more readily) by examining the surface under magnification.

A small degree of pinholing can usually be ignored, but faults are far more liable to arise and be noticed on light coloured materials than on, for example, blacks. It is also a defect which can vary considerably in extent between different batches of the same material. The pinholes need penetrate only the upper layer of coating for faults to occur.

On some materials the pinholes can increase in number and size with wear in use. They will always tend to retain dirt and attempts made to clean the covers usually only meet with partial success – the surface dirt is removed but it is not so easy to remove the dirt embedded in the pinholes. Even if this is removed the wear-enlarged pinholes cause the cover to appear soiled due to the darker "microdot" printed effect; further attempts at cleaning may cause delamination and break up the coating. In addition, the backing fabric becomes wet; it cannot dry quickly because of the coating, and moulds and mildew can develop and often show through the exterior light coloured coatings.

Faults caused by penetration of surface hairs are less common, but are relatively more so on materials with a warp knitted basecloth, or on the softer PU-coated fabrics. They tend to occur on areas of cover which are affected by both tension and abrasion during use, with breakdown occurring when the coating splits adjacent to these penetrating fibres. These splits may be so small that they are insufficiently noticeable to cause complaint, but after more prolonged use, and particularly if they are under tension, the splits may become more numerous and join to form a noticeable cracking up of the coating.

This fault is indistinguishable superficially from the faults arising from flex damage on cushions which are also accelerated by abrasion.

ABRASION DAMAGE

Several types of faults on PU-coated fabrics are due either partially or completely to the effects of abrasion, in addition to the two already mentioned. Because these materials are fashion materials, there is a tendency to select those which possess thin coatings and can become damaged during transit or use by scuffing. Normally this type of fault is easily recognisable, but it can occur in conjunction with delamination or flex cracking and is then not quite so obvious.

Abrasion can also remove photogravure printed effects from the surface of polyurethane-coated fabrics sufficiently to give rise to complaint. Covers which have suffered from the effects of hydrolysis, although few in our experience, become more prone to scuffing damage; the coatings may relax and slightly alter in surface appearance before they become damaged. The weave structure tends to show up more on the surface than it did originally – often a way to identify this particular fault.

Some abrasion faults can result if the backing fabric of a PU-coated fabric is too rigid; small puckers occur at curved seams because the material cannot easily be sufficiently deformed to enable a smooth seam to be produced and the small ridges that result can suffer abrasion damage, particularly if the underlying upholstery is somewhat hard. Puckering should be minimised by correct sewing conditions, but with one of two materials it is extremely difficult to avoid on some upholstery designs.

FLEX CRACKING

Breakdown can occur on seat or back cushions or on the arms of chairs. Sometimes areas of the front of the chair near to the base are also affected.

However, faults which arise from abrasion, and which sometimes are similar in appearance, may occur on other areas, and are usually less widespread. In general, however, this fault tends to be prominent on those parts of the furniture which tend to be flexed in use.

The coating usually breaks down in such a way that quite a large number of small cracks occur in parts of the affected area. These form a rather scaly type of appearance, particularly where the coating also tends to delaminate around the edges of the scales to expose the backing fabric. Where the coating breaks down, particularly on cushions, to give relatively large scales, approximately 9mm (3/8″) diameter, it is our experience that the resistance to flex cracking of the material has been low in unflexed and unexposed areas of the cover. This indicates that the material was not satisfactory when it was manufactured.

Sometimes the polyurethane degrades on exposure to light or, more rarely, by hydrolysis to yield a similar result. Covers, which have been exposed to light show a characteristic fluorescence pattern when they are inspected in ultra-violet light reflected from their surface (this procedure cannot be generally recommended because directly viewed, ultra-violet lamps can damage the viewer's eyesight). This inspection usually shows up the areas of cushions which have been shielded from light (e.g. areas behind buttons and inside creases formed between buttoning points) as well as showing which side of a cushion has received most exposure. If the degraded areas correspond with this exposure pattern it is reasonable to assume that the exposure has contributed to the fault. Generally, the pattern of cracking breakdown becomes much finer after the polyurethane has been degraded chemically, and the degradation often causes a decrease in the resistance to flexing.

A fine pattern of cracking sometimes occurs in which the cracks do not join up fully to separate out "scales." This is often associated with a somewhat more crinkled appearance, and, in the case of white or off-white materials, a somewhat dirty appearance.

If the coating shows fibre penetration also, then it is probable that the fault has been caused predominantly by the effects of abrasion, tension, and possibly also ageing.

Some examples of this type of fault lie in between the extremes of low resistance to flexing and abrasion. This can often be recognised as the scale edges of the cracked areas show somewhat roughened or abraded edges. The causes of all of the above faults are normally checked by testing on return cushions which we receive for investigation.

UPHOLSTERING USAGE OF POLYURETHANE-COATED FABRICS

With a fashion material of this type there are certain upholstering features which may lead to a higher incidence of returns. The following general recommendations should be observed to avoid faults. It is possible to disregard these and also to avoid faults with some materials, but it is prudent to observe them, particularly if the softer types of polyurethane-coated fabric are used.

If a predominance of white or light coloured PU-coated fabrics are used, it is preferable to install a spot check procedure for pinholed batches. This check merely consists of inspecting the surface of batches of material using a good fabric piece glass in order to find out if any of them contain a disproportionate number of pinholes in the coating.

Seams sewn across the width of the material, as it is in the roll form, may suffer from slippage, and should not generally be incorporated in a cover across the main direction of the arm cover. Particular care is needed in any seam sewn across the angle of an arm between the back cover and the armrest. Self piping should be avoided partly because, in general, PU-coated fabrics are not sufficiently resistant to abrasion, but also because the material of one of the covers adjacent to piped seams suffers damage from two coincidental sets of sewing needle holes.

Round section sewing needles should be used for constructing all seams. Sewing threads should preferably be of the core span polyester-cotton type, e.g. "Polyfils", "Koban", or should be made from cotton. Monofilament nylon threads are least suitable.

Seam puckering faults must be avoided since abrasion damage will tend to occur on the tops of the pucker marks. The backing fabrics of a few PU-coated fabrics are quite stiff and closely set, and puckering is quite difficult to avoid at sharply curved corners of panels with these materials. This is a design feature which upholstery manufacturers should consider carefully. There is a tendency to produce cushions with pleats at the corners. Undoubtedly, this may be satisfactory with some soft cushions, but again care is needed when this style of upholstery is employed.

Indeed, at all stages of design the relatively low resistance to abrasion of PU-coated fabrics should be borne in mind, and it is also prudent to consider how the furniture is packed and despatched, since damage can occur in storage or transit. In particular, areas which are subject to abrasion should not be upholstered under considerable tension, particularly if the softer PU-coated fabrics are employed.

Sensible attention to the design of furniture covered with PU-coated fabrics is a necessity. However, this alone does not ensure that faults will not be encountered. Although inspection of faults can indicate their possible causes, it is not a substitute for laboratory testing.

Dr. Weber

Farbenfabriken Bayer AG
Germany

URETHANE LIGHT-WEIGHT COATINGS AND FIBRES*

(Received January 2, 1974)

BASIC CHEMISTRY

THE FUNDAMENTAL reactions for urethanes as found by Wurtz in 1849 [1] are as follows:

$$R-NCO + HO-R \longrightarrow \underset{\text{Urethane}}{R-NH-CO-O-R}$$

$$R-NCO + H_2N-R \longrightarrow \underset{\text{Urea}}{R-NH-CO-NH-R}$$

$$2R-NCO + H_2O \longrightarrow \underset{\text{Urea}}{R-NH-CO-NH-R} + CO_2$$

$$R-NCO + R-CO-OH \longrightarrow \underset{\text{Acid amide}}{R-NH-CO-R} + CO_2$$

PU Reactions

Generally speaking the isocyanate group will react with all substances having exchangeable hydrogen atoms, ie when they are treated with Grignard's reagent they will develop methane according to Zerewitinoff [2].

It is obvious that these allophanates, ureas and acyl ureas have the possibility to react with still further isocyanate in the same manner. This shows the way to polymers consisting in principle just of isocyanates. Examples for applications of materials of this kind can be found among synthetic lacquers.

Free NCO groups tend to dimerization or trimerization, producing thermally unstable uretdion rings or the very stable isocyanurate rings.

There are further possibilities of reaction but they need not be pointed out here because these examples already show the immense variability in urethane chemistry.

It should however be pointed out that not only the possibility of a chemical reaction is of importance but also the velocity with which the given partners react.

*This paper originally presented at the Urethanes in Textiles Symposium, Manchester, England, May 1971, sponsored by The Institution of The Rubber Industry, London.

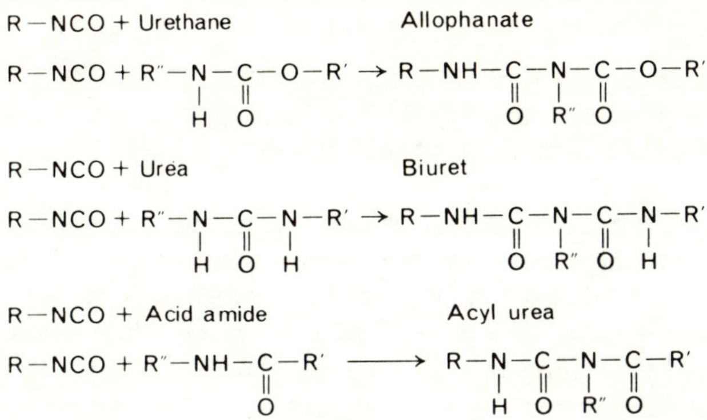

PU Reactions

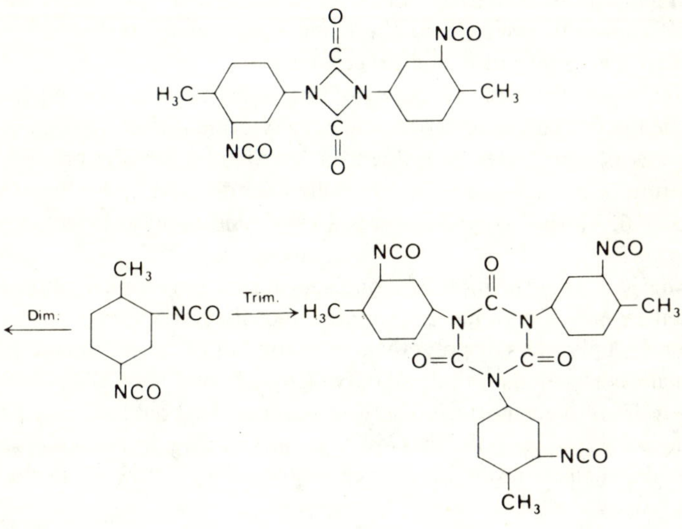

Di - tri-merisation of TDI

There is a sequence of decreasing rates of reactivity of adduct to a given isocyanate:

aliphatic amines → NH_3 → aromatic amines → aliphatic ureas → primary alcohols → secondary alcohols → water → aromatic ureas [3].

This sequence indicates the possibility that an isocyanate distributed in a blend of ethanol and water reacts primarily with the alcohol, or that isocyanate compounds react preferentially with wool although this process can be carried out in an aqueous system.

As is usual in chemistry, the presence of catalysts will influence the velocity of reactions, furthermore the relative concentrations of the partners and the temperature at which reaction takes place. In general at temperatures below 100°C the formation of urethane and urea groups is preferential, this means a chain elongation. If in the same formula only the temperature of reaction is altered to above 100°C, then the reaction turns to allophanates and biurets, thus to crosslinking the chains. This phenomenon is particularly noticeable with aromatic isocyanates [4]. Urethane and urea compounds have been found to be the more stable ones [5]. Alkaline substances such as organic metal compounds, tertiary amines, sodium alcoholate have an accelerating effect, whereas acid substances, such as lactic acid, have a pronounced inhibitory effect on the reaction.

It follows from the above that due to the multitude of possibilities of reaction and the dependence of the mechanism of the reaction of the urethanes on the environmental influences, scrupulous care must be taken to maintain absolutely pure reaction conditions and that standards once agreed upon must be strictly observed. This applies to research, to the manufacture of the technical basic materials as well as to their application and processing.

When Wurtz found the basic reactions of the isocyanates he had no knowledge of polymer chemistry. This branch of science really came into existence only in the beginning of this century, greatly influenced by Staudinger. The principle of the chemical addition of one partner to another one was followed up by Bayer when he found in 1937 [6, 7] that polyaddition is a very good method in polymer chemistry.

It means that we leave the basis of monoisocyanates and monoalcohols and turn to bifunctional or polyfunctional isocyanates and alcohols. Very soon instead of monomer alcohols polyesters or polyethers were used in the reaction and only these high molecular compounds (molecular weights between 600 and 4000 in commercial products) became technically important. It should be noted that not only synthetic polyesters or polyethers play an important part as adducts, but also natural materials, such as resins or oils, eg castor oil or tallow oil in the lacquer field.

As already stated, chain elongation or crosslinking by polyaddition calls for di-isocyanates or polyisocyanates. Only very few have gained major technical importance in general PU-chemistry, and among them are those which are of interest in textile industry: toluylene-di-isocyanate 2.4 resp. 2.6 or mixtures of both isomers, diphenyl methane-4,4-di-isocyanate hexamethylene-1,6-di-isocyanate isophorone-di-isocyanate, as well as some derivated compounds such as:

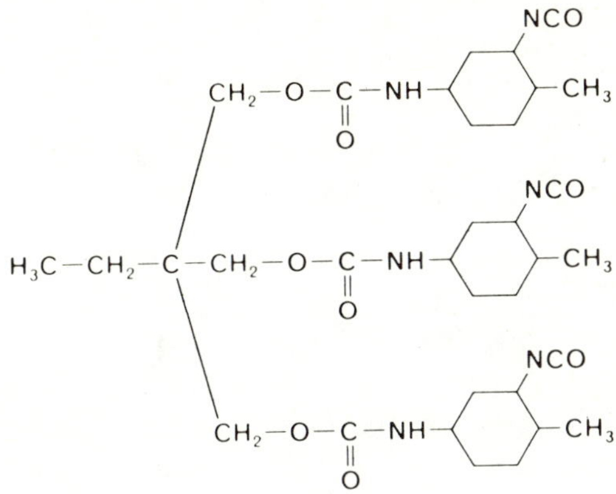

Polyisocyanates

Polyesters are derived from dicarboxylic acids such as adipic acid, phthalic acid or sebacic acid. Glycols or polyols used have a C-chain length between C_2 to C_6 or maybe even C_8. Obviously steric obstruction can be an objective in synthesis, and hence branched polyols may be introduced.

The diols used in technical production are usually derived from ethylene or propylene oxide; ethylene, diethylene, triethylene glycol etc, while 1.4 butane diol, 1.6 hexane diol and neopentyl glycol (2.2 di-methyl-propanediol-1.3) are less frequently used. Out of the range of triols and polyols, mention should be made of glycerine, trimethylol propane and hexanetriol-1,2,6 [8].

In most cases, the polyethers used in polyurethane chemistry are based on 1.2-propylene oxide or tetrahydrofurane. Polyethers of ethylene oxide have not proved successful due to their strongly pronounced hydrophilic character. Since polyethers are products of a polymerization reaction to is important whether the association takes place head-to-tail or head-to-head. Naturally branching is possible also with polyethers.

It is rather difficult to decide whether polyesters or polyethers are preferable in polyurethane coating systems. In general PU based on polyesters give coatings with a more skin-like handle whilst the ones based on polyethers tend to be a bit more rubber-like. Both types are in use.

Some tend to be polyethers, because they are resistant to hydrolysis. The chemical drawback is the possible ether-cleavage, which can be overcome by adding UV-absorbers in combination with antioxidants to the system.

HYDROLYTIC ASPECTS

Polyesters can be attacked by hydrolysis. In this respect it should be taken into consideration that there is a marked difference in the rate of hydrolysis or saponification subject to the pH-value. In the alkaline pH-range saponification is a reaction of the second order

$$\underset{\text{ester}}{R-COOR' + OH^-} \rightleftharpoons \underset{\text{anion + alcohol}}{R-COO^- + R'OH}$$

and is directly proportional to the concentration of OH-ions. In the acid range, however, the reaction takes place according to

$$R-COOR' + H_2O \rightarrow R-COOH + R'OH$$

and is directly proportional to the concentration of H^+ions due to their catalytical influence. The influence of OH-ions, by the way, is more severe than that of the H^+ions. It follows that hydrolytical effects are least in the region around pH 7 [9].

In this respect the influence of perspiration on PU is interesting. Normal perspiration is slightly acid and turns alkaline due to bacteriological decomposition.

Hole and Abbott [10] found that hydrolytic degradation of PU immersed in distilled water does not correlate with findings in actual wear trials. This could be explained by the above equations. The small quantity of acid formed by hydrolysis would not readily alter the pH of the immersion water, and hence assessments could become distorted.

It should be remembered, however, that in polyurethane chemistry things are a bit complicated, as the reactions can be easily influenced by catalysts present even in extremely low concentrations or by temperature, as stated before.

PICKING SUITABLE PUs FOR THE COATING INDUSTRY

When we look at polyesters or polyethers and their suitability for PU coatings, it must be considered that, besides the chemical constitution, the chain length, ratio of branching and molecular weight distribution will influence the properties of the resulting polyurethane to a great extent [11].

Table 1. PU Consisting of a Given Isocyanate and Different Polyesters

Polyester	T_E
Poly(hexylene-adipate-phthalate)	−18°C
Poly(-2,2,4-trimethyl–hexylene-adipate/phthalate)	− 5°C
Poly(hexylene–perhydrophthalate)	+ 6°C

A polyurethane polymer can be considered as consisting of two entirely different components: the soft segments of the polyesters or polyethers, and the

Table 2. PU Consisting of a Given Polyester (MW200) and Different Isocyanates

Polyester		T_E
	T-65	+10
Poly(ethylene-adipate/perhydrophthalate)	44	+6
	H	−3
Poly(hexylene-adipate/phthalate	T-65	−18
	H	−25

hard segments of the urethane or urea groups. Since besides a chemical linkage between molecular chains (eg branching or three dimensional arrangements) there is also physical bonding between the hard segments, the ratio of hard to soft segments is of influence on the properties of a given polyurethane. Trappe [12], for instance, showed that T_E of a polyurethane consisting of poly(ethylene/neopenthyl adipate) as soft segment plus utane-diol 1.4 and TDI as hard segments, rose from −25°C to 0°C when the ratio of the hard segment was increased from 17 to 0% by weight. Furthermore, steric hinderance will cause this physical bonding to be impaired. It also becomes apparent that the molecular weight of the polyester used will influence the properties of a polyurethane. This can be shown by determining T_E (Glass temperature) [13].

Table 3. Influence of MW of Given Polyesters on Glass Temperature T_E

Polyester	MW	T_E
Poly(ethylene-adipate/phthalate) idem	2000	+18°C
	1000	+25°C
Poly(hexylene-adipate/perhydrophthalate)	2000	−23°C
idem	1000	+2°C

In textile coating it is always of advantage when T_E is as low as possible because then a softer handle of a coated fabric can be achieved. The variety of possibilities in PU-chemistry manifests itself in different ways.

Because it is possible to synthesize thermoplastic, elastomeric as well as poromeric PU, and this perhaps in several steps there are quite a few different ways of application that are practicable. These again may call for specially designed raw materials for the form of application intended and, finally, by the purpose for which the material is made for.

PUs had their first major outlet in the foam industry. We can say, in a simplified manner, that polyesters, (poly)-isocyanates, water and catalysts were made to react,

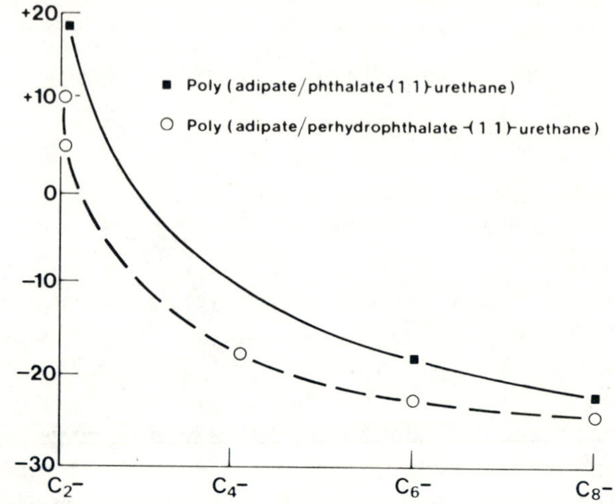

Relationship between glass temperature & C-chain-length in the glycol of polyester (MW 2000)

Figure 1.

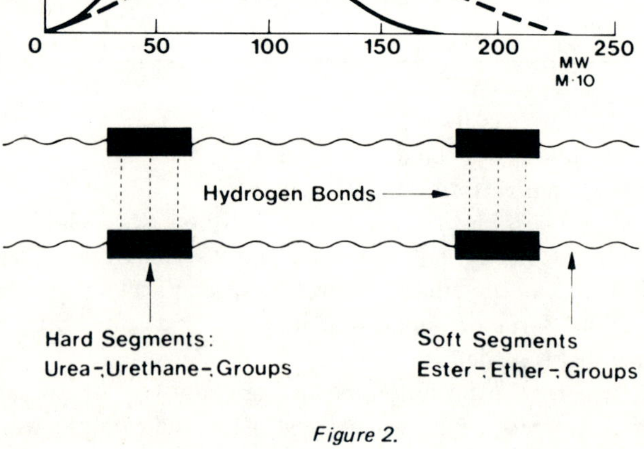

Figure 2.

resulting in a porous and elastic foam.

Although attempts had already been made quite early, a foam of this kind could not be produced on a bulk scale direct on a given textile support. The reasons are obvious. Because "direct coating" techniques were used, the relatively low molecular weight reaction compound penetrated too far into the textile support, thus producing a boardy material. Furthermore, the foam formed was too thick to be of interest, seen from the textile expert's point of view. It was also a handicap that a surface skin was formed or two skins formed in the supports when a sandwich construction was chosen.

Today this technique is no longer rejected as it used to be, but it is now being used outside the textile trade. In construction engineering preformed parts can consist of a sandwich construction containing a PU which was allowed to penetrate a coarse fibre non-woven whilst the foam formed.

TWO-COMPONENT PUs IN THE TEXTILE INDUSTRY

Instead of using polyesters and isocyanates it was decided to go in for prepolymers and polyisocyanates in textile applications. Polyesters were chain-extended with TDI and were supplied in the form of chips or solutions in ethyl acetate. Incidentally, one of the first applications intended was for chintz-materials. Soon the advantages and the flaws of PU became apparent and a rapidly growing textile coating industry based on these two components began to spread.

The method of application using direct coating methods is simple. The breakthrough of PU-coated textiles is connected with the idea of choosing raised cotton fabrics for the textile supports. This construction gives an interlayer between the relatively thin PU film and the base fabric and due to this construction a soft handle, textile draping qualities etc can be achieved with lightweight and yet tough coated materials.

Obviously a lot of skill has been used trying to improve the materials in question. Many kinds of fabric construction, fabric finishes, ways of raising, cropping etc have been evaluated to set the standards of today's production. Furthermore, successful attempts have been made to coat raised knitted fabrics. Fibres or filaments used in these materials are polyamides and rayon materials. Base materials consisting of other synthetic fibres both in knitted and in woven structures are being developed. Coating knitted fabrics by the direct method is rather difficult. The transfer method is such more advisable. It took some time to develop the types of release papers needed for these purposes, but nowadays they are available in sufficient quantities.

PU foam prompted industry to try fabric foam sandwich constructions. The laminates have a big market. PU or acrylics were used as adhesives. Foam laminates gained importance and by this technique the foam thickness applied to the textile vanished to almost nothing. Now bonded articles dominate this field. Incidentally adhesive bonding regains importance. It is interesting to point out that PUs used for adhesive bonding and for textile coating belong to the same category or even the same range.

PU two-component systems require some time for crosslinking. Although the addition of water or alcohol to isocyanate is an exothermic process, production in the coating field cannot rely entirely on this mechanism. Catalysts and releasing agents are therefore incorporated in the coating mix. It is important not to use catalysts that will primarily increase the velocity of crosslinking but will on the other hand reduce the resistance of the PU to hydrolysis considerably, eg stannous compounds is to be considered unsuitable.

Bayer [18] and others [15] showed the crosslinking in the systems in question takes place only in the presence of catalytical quantities of water. Furthermore,a minimum temperature for crosslinking is essential, as follows from known thermodynamic reasoning. In this case the minimum temperature ranges between 15–20°C and the catalytical quantities of essential water will be available when the relative humidity exceeds 40% at 20°C, and when the necessary pick-up is ensured.

It stands to reason that the PU two component systems have a disadvantage in plant scale coating, because the result of the production can be controlled only after a certain time has elapsed.

ONE-COMPONENT PUs IN THE TEXTILE INDUSTRY

As a consequence, one component PU of the thermoplastic type dissolved in MEK/DMF blends gained major importance. These materials belong chemically also to the PU prepolymers. Only they have been chain extended to a high degree. They are still soluble in a reasonable percentage in these solvents.

The application of these types of PUs calls for modifications in the coating machinery set-up and other equipment. Coating should not be direct but only by the transfer method. In the case of direct coating the coating mix will tend to become less viscous when the material is heated up in the drying unit. Thus the mix penetrates the basic material and the handle of the coated fabric will be less textile.

Working with the transfer method calls for adequate release materials – release papers or release textiles. Only release papers have so far gained technical importance and are readily available. It may be of interest to note that so far no scientific method has been published according to which it can be predicted whether a given release material in use in plant scale production has to be discarded after a couple of runs or whether it will still stand up to one or more runs. A test method of this kind would greatly improve the standardization of plant production and would cut production costs.

The technical method of transfer coating is nowadays to be considered a common art also with regard to the PUs. But it must be borne in mind that the solvent system used for the top coat should also be the solvent system for the adhesive coat. This will also hold true when a one component top coat is laminated to the base fabric by a two component PU. It is of course another matter when an aqueous emulsions-based bonding agent is used as the adhesive coat.

Once the one component PUs in the solvent system DMF/MEK had become available, the next step in development was to try to find PUs that are soluble in other solvent systems. This is mainly governed by considerations of toxicity of DMF/MEK and because of possible contamination of air or water. In an experimental stage PUs dissolved in toluene/isopropanol are being evaluated. It is not to be considered as an improvement in a direct way, but indirectly because the solids content exceeds 50%, and the solvent amount is therefore reduced considerably.

One of the major advantages of the one component PUs to be considered is the fact that production can be controlled definitely after the coated textile has been separated from the release paper. Also the possibility of heat sealing should not be disregarded. Nevertheless a crosslinked PU would still offer some other advantages, as is well known.

It stood to reason therefore to develop PU systems that are one component compounds to the coater and, from a chemical point of view, two component PUs. A compound of this type would be a prepolymer of the type used in the two component systems blended with a formaldehyde compound. This mixture is stable under normal storage conditions and will be activated only under elevated temperature conditions.

No doubt, some PU systems will be available in the future where not a blend but a homogenous, stable and yet reactive compound can be applied. Chemists never give up. In this respect mention should be made of systems where isocyanates blocked with oximes were incorporated in the coating mix. Up to now these systems failed to prove a success. When de-blocked, the oxime is set free, and coated textiles with a crater-like surface are frequently the result, due to evaporating oxime also after a surface skin has formed on the PU coat while running through the oven.

In the chain of developments another possibility was looked into very intensively. As was pointed out before, PUs with very hydrophilic behaviour can be synthesized. As a sideline one could mention that there is a polyurethane which acts as an emulsifier and a thickener in aqueous textile printing pastes.

PU AQUEOUS SYSTEMS

PUs having an ionic centre such as

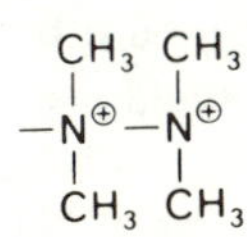

are available as aqueous emulsions. It is especially interesting to note that these emulsions can be prepared containing no emulsifiers. Because of the ionic centres present in the molecule it is sometimes inevitable to crosslink these products with formaldehyde donors. Of course due to the skill applied in the synthesis of these materials the necessity of crosslinking is the exception. One can find a compromise between stability of an emulsion free of detergents and a resulting coating which is still sufficiently stable to attack by moisture.

With these aqueous emulsions another step in the chain of development has been taken. Water presents no hazard to security in production. The coating mix in general has a long shelf life, and this is due to the fact that the PU is one component product.

Coating and laminating are two of the possible outlets for these types of PUs. Coating is done mainly by the transfer process. Here a handicap was encountered. Release papers had to be – and have been in the meantime – developed that could stand up to these aqueous systems.

If one considers the chain of steps in the PUs developed for coating textiles it becomes obvious that laminating a preformed PU film to a given textile would be of interest. This has been done, both with microporous and with homogeneous films. Elastomeric technique, the substrates ranging from real split leather, woven and knitted fabrics to non-woven fabrics.

COATING BY LAMINATION OF FOILS

The films can be produced on a melt calandering machine and applied in the way known in the trade. Only if one considers lightweight coated textiles one is a little bit off the track. These methods apparently will have their major outlet in the heavy duty field.

Mention shall also be made of a way for coating carpet backs with PU systems that are synthesized from polyethers or polyesters and polyisocyanate *in situ* on the carpet back. This is a very interesting and promising way of applying PU in the heavy duty field. Here one of the advantages of PU is essential for the economy of the procedure: PU can be filled to a very high degree with colour pigments and with fillers. Due to this phenomenon a heavy weight coating is achieved with a comparatively thin layer. This is exactly what people want, when textile tiles are used in floor coverings and especially in the do-it-yourself method.

GENERAL DISCUSSION OF PU QUALIFICATIONS

The sequence of PUs used in and developed for the field of coating and laminating textiles can only be understood if there are a few advantages to be found in these polymers. If so, then they would also stand chances in other branches of the textile industry.

The main and characteristic properties of PUs can be listed as follows:

They can be tailored to fit a variety of applications, both from a chemical point of view and from the physical appearance.

They can be made to meet the requirement of flexibility at low temperatures and yet not alter the outward appearance at elevated temperatures. PUs don't need the help of plasticizers to meet this requirement.

They can be dry cleaned and because no plasticizer will be extracted in this process, the handle will not change.

PUs used for textile coating have a high resistance to abrasion and this is the reason for their use in lightweight textile coatings. The coat applied is relatively thin (eg 70 g/m^2 solids) and yet stands up to normal wear.

They have a very good adhesion to base materials used in the coating or laminating field. This becomes apparent also by the fact that they can be used in special modifications as primers for other polymers. Cohesion properties are also extremely good.

They can be coloured in a wide range with suitable pigments, both in uniform shades and patterned (printed).

They adsorb and desorb water. Water vapor will be able to migrate through even a homogeneous PU layer. This phenomenon is used to advantage in the artificial leather industry (garment application) and especially by means of microporous PU structures in synthetic leathers in the shoe trade.

They can be synthesized in such a way that they will behave as elastomers, plastomers or duromers. In this respect one should remember that at temperatures above 150°C the physical intramolecular bonds in the PU macromolecules begin to break up and will rearrange themselves on cooling down. The elastomeric character of PUs thus relies to a great deal not only on chemical basis but also on the physical arrangement of the hard segments of the macromolecules. It goes without saying that there is no material which has only advantages. The possible disadvantages could be summarized as follows:

PUs derived from polyesters will be subject to hydrolysis. As a remedy, select glycols with a C-chain length above 4, and increase the ratio of urea segments and urethane segments in the macromolecule.

PUs derived from polyethers tend to ether cleavage under the influence of UV light and elevated temperatures. As a remedy: UV absorbers and antioxidants. The combination of UV absorbers and antioxidants may well show a synergistic effect, depending on the system used.

PUs based on TDI tend to yellow under the influence of light. Remedy: TiO_2 of the rutile type stabilizes very well. Use PU coating mixes only in a pigmented form.

TEXTILE APPLICATIONS APART FROM COATING

If all these properties are added up one should expect PUs to find applications in the textile field apart from coating and laminating. This holds true, as was already pointed out in a few previous remarks.

Wool can be treated with PUs to become resistant to felting. Application of the PU can be from an organic solvent system or possibly – but not established in plant production – in an aqueous system although the PU in question has free isocyanate end groups. Here it might be mentioned that isocyanates may react with the fibres of the basic material. This was an assumption as of old, but it can be considered to have been proven by now [17].

Solid PU chips of a thermoplastic character stimulate the wish to apply them as heat sealing agents. Ground to fine powders they have been applied for this purpose on an experimental basis. They have undergone experimental runs by the powder point technique, they have been printed in the form of pastes and have been applied in the common way of sieving them on the textile base, all yielding appreciable results.

The elastomeric character of PUs predestines them as a chemical basis of elastomer fibres. Chemically such a fibre could be as shown below. In most cases the multiple filaments stick together in bundles. They have a distinct textile character, due also to the surface which is rough when looking through a microscope. PU fibres have an amorphous structure when relaxed and show marked crystallinity when stretched. This can be shown by X-ray diagrams. It is assumed that the unorientated

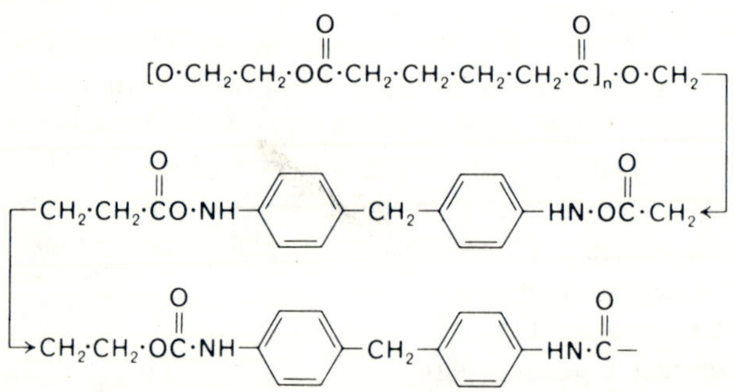

Linear Elastomer PU fibre

soft segments (polyether segments) become orientated and crystallize while subjected to stretching. The crystallization is only a temporary one and breaks down when the fibre is allowed to relax.

PU fibres can be dyed just as textile fibres are. All processing in the dye shop is done in such a way that the specific chemical constitution of these fibres is considered. Suitable dyes are among the acid and 1:2-metal complex dyestuff ranges.

It is not so easy to process elastomer fibres, and special new machinery had to be constructed to cover the mechanical aspects. Elastomer fibres are used generally only in blends in the form of covered yarns or core yarns. They compete with the rubber yarns and are to be found in some stretch articles. I may mention cloth for sportswear, elastic medical bandages, hosiery and subhose [19].

MECHANICAL APPLICATION OF COATINGS

Returning to the field of coating and laminating, it became apparent that four methods of application seem to be most important: direct coating, transfer coating, laminating of prefabricated films, and melt calandering.

Turning to coating with a doctor blade two techniques are in use, each of which offers advantages and disadvantages.

Today one can assume that there is no big difference in adhesion due to application technique. Perhaps direct coating can be considered to be simpler to run than transfer coating. Guiding the material in the direct method is less complicated, because there no two different materials need be guided. Especially the guiding of paper calls for extreme care.

In transfer coating more solid content per coat can be applied. The viscosity of the coating mix has to be controlled more accurately in direct coating as in transfer coating, because in transfer coating unwanted penetration into the substrate cannot occur. Transfer coating in general tends to give a softer handle to the coated textile.

It is obvious that most of the animosity against transfer coating and the "ifs" and "buts" have been overcome by the developments in the past years.

There remains the question whether direct coating is obsolete or not. This can be decided upon only by industry itself, but one point of view may well be discussed here. In a modern line direct coating can be done with a speed of about 25 m/minute. Transfer coating can run safely in production with about 15 m/minute.

A coated textile for apparel use would be made by both techniques and the production schemes could be as follows:

	Direct	Transfer
1.	Pre coat, clear	Top coat, pigmented
2.	Coat, pigmented	Adhesive coat, clear
3.	Top coat, pigmented	
4.	Finish	Finish

Production schemes for PU-coated textiles

The solid add-on would amount to about the same, because in transfer coating more solids can be applied in one stroke.

Running a plant, one could assume that in direct coating a coating equipment for the pre-coat, a tandem unit for the pigmented coats and a machine for the finish would make a good working unit. Taking the transfer system one could get along with a set-up as before, but omitting the unit for the pre-coat.

Considering these assumptions the machinery installation necessary for transfer coating needs less floor space. The costs for the equipment cannot be commented upon in general terms; they have to be evaluated individually. One should not disregard as a cost factor, that in direct coating a needle-loom at least for the pre-coat and preferably also for the first pigmented coat is advisable.

The cost of the machine-hour (including wages, raw-materials etc) would have to be distributed in direct coating to about 30 m^2/minute versus 18 m^2/minute, in transfer coating if a plant production rate of 80% is assumed in both techniques.

In transfer coating one cost-causing element has to be taken into account on top of what is needed in direct coating: the release carrier, the price of which cannot be neglected. Reference is made to previous remarks. It follows from this that direct coating is by no means obsolete and that one should consider both ways of application as complementary to each other.

DISCUSSION OF PU COATED TEXTILES

Let us look at a few samples to demonstrate the applied theory.

A sun shade coated on one side uniformly with a PU and on the other side in different coloured stripes shows very well that with only one coat and a very low add-on good coverage can be achieved. The technique used is direct coating. Coating of the stripes requires a doctor blade divided into different sections and having receptacles for each colour.

A light rainwear material is also coated directly. The film applied is practically scratched on the polyamide material, a continuous layer is not desirable. As the coating generally is worn on the inside of the raincoat, a clear coat is usually applied.

Fishermen would not be satisfied with a lightweight raincoat, but would rather have slightly heavier material. For safety reasons a brilliant colour such as a bright yellow, which is noticeable even from a greater distance, is preferred. Here also direct coating was used, although the transfer method may well be applied.

When coated materials are to be worn actually as garments the above mentioned materials have not been accepted by the public. It was necessary to find a better build-up of a coated lightweight material. This was done by way for placing a go-between between the PU coat and the base fabric. Two methods are known to achieve this effect: a layer of foam or the pile of a raised fabric. The latter way was chosen, when PU coatings got their break-through in the apparel and upholstery trade during the last few years.

Using a woven structure of a pile fabric, originally only the direct coating technique was applied, but to-day with one component PUs pushing forward the transfer method is gaining major importance. This applies especially for upholstery materials, where heavier coating is desired as well as the possibility of welding or heat sealing the material. Such articles, on the other hand, are coated only by the direct method. Aqueous PU dispersions will be mainly applied by the transfer method.

Looking at coated knitted fabrics one would recommend the transfer coating method. Knitted goods have had a hard time in securing a market share. This is probably due to different circumstances. Firstly it is difficult to construct a lightweight, easy-to-raise and yet stable knitted fabric and, secondly, the public was used to the tough outward appearance of coated cotton fabrics. Again it should be pointed out that this type of material, be it a coated knitted or woven fabric, is basically influenced in quality and appearance by the quality of the basic material as well as by the skill with which the coat was applied. It also stands to reason, that one has to evaluate very precisely whether a given material will stand up to the demands governing a special field of application. This demand holds true for the selected base fabric as well as for the PU applied. The difference between a garment and an upholstery material is, beside the heavier quality used in upholstery also the fact that one expects a longer wear life of these materials. Non-woven materials should always be coated by the transfer method or by laminating a film to them.

Turning to tarpaulins for heavy duty, used in the transportation trade, it becomes apparent that PUs applied by the coating techniques don't stand a big chance, production costs being too high. Here melt-calandered coats have by far a better position. It will have to be shown as yet whether PUs applied by this technique will be able to gain a share in the market. Heavyweight PU carpet backings are applied by spray coating.

Concluding it may be pointed out, that finishes of all kind such as high gloss, semi-gloss, dull appearance, printed pattern, irregular designs etc based on PUs as a carrier, are common in different fields including the leather and paper trade outside the textile coating field. Application techniques may be spraying, screen or roller printing.

REFERENCES

1. A. Wurtz, *A Chim* (3) 42, 43.
2. Tschugaeff-Zerewitinoff, *Ber* **40** (1907), p. 2023.
3. J. H. Saunders and K. C. Frisch, "Polyurethanes, Chemistry and Technology," New York, Interscience Publ (1962, 1963).
4. H. Oertel, "Ullmann Enzyklopaedie der Technischen Chemie," 3rd ed, Vol. 14, p. 340.
5. Wagner, Mennicken: "Aliphatische Isocyanate als Rohstoffe zur Herstellung von Lackfilmen," Papers by Bayer on the Fatipec Congress (1962).
6. DRP 728 981, I. G. Farben Ind. O. Bayer.
7. O. Bayer, *Angew Chem* 9, 257–272.
8. H. Nordt: Kunststoff-Handbuch, "Polyurethane," Vol. VII, p. 48.
9. Eucken-Wicke, *Grundriss d Phys Chemie,* 8. Aufl, p. 416.
10. Hole and Abbott, Satra-Symposium, Blackpool (April 1971), L 10.
11. Koch, *Mell. Text. Ber.* 11/51 (1970), 1313–1317.
12. Trappe, "Advances in Polyurethane Technology, London (1968), p. 106.
13. Koch, ibid.
14. O. Bayer, *Angew Chem* 62 (1950), 3, p. 64.
15. Berger, Der Einfluβ der Luftfeuchtigkeit auf die Reaktion von Zweikomponenten, "Polyurethan-Lacken, Papers by Bayer" on the Fatipec Congress (1962).
16. Dieterich, *Keberle u. Wuest, I. oil. Col Chem Assoc* 53 (1970), 363–379.
17. Rose, not yet published.
18. P. A. Koch, "Faserstoff-Tabellen PU-Elastomerfasern," ed August 1968, DK 677.494.66.
19. Pullen, *Chemiefasern* (1967), 1; and Avellis, Vitz, ibid (1967), 2.

J. E. Bell and D. J. Hanrahan

MACHINE DEVELOPMENTS FOR POLYURETHANE-COATED FABRICS

(Received January 2, 1974)

INTRODUCTION

MACHINE DESIGN IS directly related to chemical process and product development. Very few mechanical revolutions have occurred in process machinery since the early days of nitrocellulose, although there have been many refinements introduced, e.g., forced convection air heat treatment and precision gap controlled coating.

The 1920s saw the introduction of nitrocellulose. The coatings were solvent based solutions applied by the direct spreading technique to woven cotton fabrics and paper. Process temperatures were low, therefore, steam was the heat source with air for solvent vapour removal.

There has been considerable development of machinery to handle this type of solvent coating where convection air ovens replaced steam platens and oil, gas or electricity can be used as the heat source.

The 1940s saw PVC appear with plasticizers replacing solvent as the vehicle. Higher process temperatures were required to cure or fuse the PVC, electrical heating was introduced in the form of infra-red lamps or elements. The direct spreading technique continued, but flameproof equipment became unnecessary. The advent of expanded PVC coatings placed new demands on the machine builders as this product contained temperature sensitive blowing agents requiring very accurate coating and equally very accurate and uniform heat treatment, e.g., one 0.001 inch coating error could mean 0.003 inch error after expansion and curing.

A new manufacturing technique using release or transfer papers was developed to act as a carrier for the PVC coatings permitting knitted textile materials to be combined successfully.

A machine was required to produce an expanded vinyl in one operation and this is done today using an inline multi-coating unit machine. Many machines made in the last six to seven years have as many as three coating and heat treatment sections and some have four or more.

The 1960s saw the first serious commercial development of polyurethane as a coating. The new material received phenomenal acclaim throughout the coating industry, despite solvent base solutions. The machinery and processing techniques

*This paper originally presented at the Urethanes in Textiles Symposium, Manchester, England, May 1971, sponsored by The Institution of The Rubber Industry, London.

developed for expanded vinyl are in many ways ideal for polyurethane coatings, although new problems have appeared taxing the technical resources of the chemist, engineers, and everyone associated with producing the material.

Urethane vs Expanded PVC

Examine now at the differences between PU coated fabrics and expanded vinyl coated fabrics. Both have a top skin protecting a cushion layer supported by a textile material. The cushion layer is either a textile pile raised from the textile base fabric, or a cellular vinyl layer. Polyurethane skin layers are usually 0.002 inch to 0.003 inch total thickness and expanded vinyl skins around 0.010 inch. PU film thickness is restricted at the present time due in the main to:

(1) Low solids content of solution, usually between 20% and 40% compared to PVC/plasticizer.
(2) High cost of polymer and solvents compared to PVC/plasticizer.
(3) Higher production costs.

Polyurethane coatings appear to have marked advantages over expanded vinyls for certain application, they are lightweight, have good drape qualities, are dry cleanable, possess stability at high and low temperatures with extremely good abrasion resistance. One also hears reference to air permeability with certain types of construction, but above all it is the aesthetic appeal that attracts the customer.

Successful manufacture of a polyurethane-coated fabric depends on the machine being capable of handling all the correct ingredients in a proper manner.

The machine will usually comprise take-off and take-up units for the fabric or transfer paper, a coating unit having either a floating knife, knife over rubber blanket, knife over roll (rubber or steel) or a combination of all these features, a dryer and cooling cylinders. These units may repeat themselves according to production technique.

Given a good quality pile fabric with an even pile, the solution can best be applied directly using the floating knife technique. This method, correctly applied, prevents penetration of the solution too deeply into the fabric, but at the same time develops a good bond with the pile or fibre tips. Additional coats are sometimes applied before top coating. This method of coating is critical, as coating weight is directly related to fabric tension. Too high a tension can cause penetration and loss in width, too loose a tension increases the coating weight. It is for these reasons that very close woven cotton and nylons are the most suitable materials for coating by this method. The knife shape used is generally a sharp profile set almost vertically.

The air-knife-rubber blanket coater is often used to apply a secondary colour on PVC embossed surfaces. The colour is deposited in the valleys of the design by the scraping action, leaving the top clean. The top of the design must be smooth and of a shallow nature. This is often called in-filling or antiquing.

After application of the primer coats, the fabric is often calendered to soften the product. The top coat can now be applied usually by the rubber blanket coater, because only a small amount of tension can be tolerated in the coating zone.

The rubber blanket resilience also allows irregularities to pass through the coating zone, especially when lipped coating knives are used and the blanket correctly tensioned. A centre support roller is often fitted to provide additional blanket tension. If the knife position is on top of the centre support roller, the blanket coater .can closely resemble a knife over rubber roll coater. Additional coatings are usually applied in this manner until the desired coating weights – handle – permeability are achieved.

As you can see, direct coating methods are critical, require the very best quality materials, are labour and time consuming. Other products can be made using this technique, for example, sail cloth, foul weather clothing (cotton or nylon based), tarpaulins.

However, if full use is to be made of all the recent development in the coating fabric industry, then the transfer coating system is probably the best method available for producing PU coated materials.

The two-head coating machine is relatively simple. The release or transfer paper is threaded through the plant and taken up on the paper winder. A polyurethane solution is applied to the paper at the first coating station, usually by a profiled knife over steel roll unit. Solvent is removed from the solution in a specially designed drying oven. From this point, the coated paper passes through a surface cooling zone and is then given a second coating. A textile is then combined pile down on to the coating before being dried again and cooled.

The sandwich of paper, PU and textile may be wound together or independently according to the type of coating employed. In either case the material is taken up under tension controlled conditions. Whilst this brief description of the process and machine may make it appear simple, there are many problems to be overcome, closely affecting both chemist and engineer. The temperature settings for the oven zone – solvent balance, speed, coating weight – are but a few of the problems that must be resolved either on the plant or in the laboratory. A lot of work has already been done, but much more will be done over the coming years to refine the process and fully develop products.

I would now like to look at transfer paper as this is the base on which the majority of PU coated fabrics will be made in the future and this has considerable influence on machine design.

The transfer paper is often a silicone or similar type of coated paper specially prepared for the coating industry. There are qualities to suit the full range of vinyls as well as PU coatings – plain, embossed and glossy. These papers by their special nature are expensive and should be treated with respect from all who handle them. Damage to the paper before processing will lead to plant stoppage and lost production. The paper should be stored off the floor to avoid deformation. A "D" shaped reel will not take-off evenly causing the reel to snatch and surge. Tension variation develops at the coating knife causing marks and uneven coatings. Edge cuts or tears can cause chaos at the coating knife. Similarly, a torpedoed reel will not run true causing tight and slack sides with inevitable creases and tears. If this high quality paper is to run successfully the machine must be designed and built correctly. This means all the rollers, cylinders, take-off and take-up shafts must be accurately aligned. All these rollers must be precision machined with negligible eccentricity and should be balanced.

Adjusting rollers can be placed at critical points to accommodate difficult reels of paper. Damaged and used paper should always be inspected and checked, and accurately rewound on a special independent machine. If attention is paid to this part of the process there is a very good chance that the rest of the process will go smoothly.

I would like to mention at this time how important it is to treat the solution in a similar manner to the paper, in particular it should be kept clean and free from dirt, in a healthy atmosphere, and checked or filtered before it is finally used on the coating machine. It is rather frustrating to say the least to have a breakout on the coating machined caused because the solution contains foreign matter. No amount of ingenuity on the part of the engineer or machine builder can cure or resolve problems stemming from dirty material and faulty paper. To have two or three coating knives, one after the other is not the answer as it is only a matter of time before the second and the third become polluted.

We will now examine coating units. Precision knife over steel roll coating units are the most commonly used in transfer coating. The system was developed primarily for coating paper and fully developed for coating PVC plastisols on to transfer papers. The machine can apply extremely accurately coatings with reliability and consistency. For example vinyl coatings in the order of 0.003 inch and PU coatings in the order of 0.001 inch thickness can be made.

As with other roll coating systems the wrap of the paper on the roll should be not less than 90°, excellent results being obtained where the wrap is between 140° and 180°.

Vertical adjustment of the coating knife is a precise function using precision manual or auto adjusters. Pneumatic quick lift on the knife can release lumps, splices or torn edges. Beta gauge control is fitted to many machines making them fully automatic. The knife can be set at any angle or completely revolved to use a different knife profile. Alignment of the knife with the steel roll horizontal axis is vital, otherwise wedge shaped coatings are produced. The knife is usually bedded into the steel roll to avoid irregular cross sectional coatings.

The precision knife over a steel roll coater can be extended by having an air knife attachment fitted to allow direct coating of woven fabrics.

Sharp edge or small radius blades produce thin coatings. As the blade base is increased and lips are added a compression zone is formed causing the coating thickness to increase. Any solids or foreign matter in the solution tends to follow the rolling bank and are kept out of the coating zone. If the solution in the bank is allowed to fall very low then the particles pass into the coating zone causing streaks and trouble.

Lip knives could be said to serve three purposes:

(1) They increase the coating area thus increase the wedge action which influences the coating weight.
(2) They give a sharp cut off after the metering zone.
(3) They make a ledge on which lumps/solids collect (applies mainly to PVC).

The three or four roll reverse roll coater is generally used for high speed coating of thin films on paper bases and enjoys tremendous popularity in the USA. There are two very important differences between this method of coating and those of knives. The coating is applied by a roller and this roller always revolves in the opposite direction to material travel.

The metering roller or rollers by definition meter or case the solution on to the applicator roller. Thus the coating thickness has been formed before it reaches the material. Applicator/metering roller eccentricity and parallelism are within ±0.0002 inch. The solution is contained between the applicator and the metering rolls. The coating weight can be adjusted whilst the plant is running by increasing or decreasing the speed of the applicator roller and gap between these rollers. Coating can be started and stopped instantly by dropping the backing roll away.

I have been describing the main features of the three roll reverse roll coater. The four roll machine introduces a pick-up roller and is usually employed on lacquer coatings or very thin solution coatings.

The main advantages of the roll coaters are:

(1) Very accurate films are formed.
(2) Suitable for high speed coating – 100 feet per minute plus.
(3) Instant on/off coating – joins – damaged paper edges pass through.
(4) Little or no trouble with dirt and streaky coatings.

One immediately feels that this machine should have completely overtaken the knife coating unit, particularly when our manufacturing costs are continuously rising. However, where solvent solutions are concerned, and particularly polyurethane and two part systems at that, there are three good reasons for their lack of popularity.

(1) Unsuitable for direct coating of plain woven or pile fabrics.
(2) Extremely difficult to clean as there is a tendency for the isocyanate to bond to the metal applicator and metering roll, to say nothing of the effects the solution has on the rubber backing roller.
(3) Solvent loss in the bank and on the applicator roller will lead to difficulties in coating.
(4) Two to three times more expensive than a precision knife coater.

Just as it is important to have an accurate method of coating and applying the polyurethane solution to the paper or fabric, so it is of equal, if not greater, importance to have a good means of combining or laminating the textile fabric to the adhesive layer. At this particular point the product can be made either beautifully soft and flexible or rough and hard resembling tough paper.

The object of the laminator is to combine the raised or pile fabric – pile down on to the adhesive coat in such a way that adequate bond strength is obtained to meet the end product specification yet retain the flexibility and handle of the base fabric.

I propose to discuss one basic system adding that there are as many different designs of laminator as there are colours in the rainbow, each offering a particular extra feature more desirable and so on.

A typical laminator would handle large reels of material, say 40–50 inches in diameter. The reeling shafts are very free yet provided with a sensitive and controllable disc brake. It is very important that the unreeling or take-off shaft should not bend as this will create problems at the laminator. There are lead rolls, draw rolls and width control rolls, necessary if tension in the material is to be controlled accurately and if width loss is to be avoided. The laminating roll should not be less than 6 inches diameter. This roll is fitted with fine adjustment so it can act as a pressure – gap or floating roll. The laminator as a whole is adjustable relative to the coater, a very important feature. The minimum distance is usually 18 inches up to a maximum of 48 inches. This range of adjustment will allow the processor to find the optimum tack point for a given solution and fabric thus ensuring a good bond – flexibility ratio.

A drive to the unit provides accurate tension control. The laminating roller can be adjusted independently side to side in the same way as a coating knife. We are after all, handling a product which is as accurate as the coating and should be treated as such.

DRYING, SOLVENT EVAPORATION, OVENS

The dryer must remove the solvent efficiently, especially when one is handling expensive raw materials. A difference in heat demand of one million BTUs per hour between one machine and another makes a big difference when it come to costing out the product. The most efficient dryer is one designed around the principle of forced convection air drying. The whole subject of drying and in particular the drying or removal of solvent could form a paper in its own right and, therefore, I do not propose to go too deeply into the effects of boiling points, lower explosive limits, heat transfer rates etc.

Considering we are removing solvent from the coating, there are certain precautions to be taken.

(1) Some equipment will have to be flameproof.
(2) The air solvent mixture must be at a safe level.
(3) Careful selection of the heating system is necessary.

In the case of polyurethane coatings, we can be faced with a one or two part system where solvents may vary between ethyl acetate to DMF with boiling points ranging from 68°C to 153°C.

You may be restricted in temperature to 100 or 120°C when using certain nylon materials. This immediately limits the evaporation rate of DMF. We can easily apply and transfer heat to the coating, but this could be disastrous with THF and ethyl acetate unless there is a way of controlling the heat transfer rate.

The first dryers were very simple because the solutions were two part systems using ethyl acetate as the main solvent. All that was required was an exhaust hood causing a current of air to pass over the sheet.

A little gentle back heat from a steam platen could be useful. Unfortunately, the product made with this system created other problems resulting in the introduction of a one part system with a higher boiling point solvent.

Therefore, a machine is required to handle:

(1) High/low boiling solvents.
(2) Have infinite variable air velocity.
(3) A large temperature range.
(4) Infinite variable auto temperature control.
(5) Comply with the Factory Inspectorate regulations.

A typical machine that fulfils these requirements would have high performance forced convection dryers in the order of 40 feet in length, usually divided into three separate temperature zones. Each zone is fitted with a fan, heater unit, plenum chambers and filters. They can handle high/low boiling solvents because the heat transfer rate can be varied from 0 to maximum. This adjustment can be made to each zone. The temperature range is between 40 up to 200°C or more if required. An exhaust fan is required to ensure that the machine is effectively scavanged of vapour laden air. The system is designed to suit the process condition most likely to be met by the particular customer in question. When variation of process conditions factory to factory reduces, then standard dryers are feasible. However, until that day dawns, it is better to have the dryer designed to meet specific conditions.

Air is heated and discharged into the plenum chambers by a fan. The air impinges in fine streams on to the coated material, evaporating the solvent. The solvent laden air is withdrawn uniformly from the enclosure by the exhaust fan. There is a very careful balance of fresh air and exhaust air so that the solvent vapour content is maintained at 25% by volume of the lower explosive limit of the solvents used. The machine can be made to have complete sectional evacuation of air or, if necessary, the air can be re-circulated.

When a machine is made with several variable controls, gauges are fitted to monitor air flow, thereby helping the operator to adjust the machine to suit process conditions. These gauges also clearly indicate when the air flow is being restricted by, for example, blocking filters. Flow detectors are fitted so that if there is no air flow the machine automatically stops and cannot be started until the fault is corrected. The inclusion of filters in this type of machine cannot be over-emphasized and are required on all air inlet points as well as interior air circuits.

I would like at this stage to make clear the position of lower explosive limit with relation to dryers and solvent evaporation. It is essential to calculate out the exact solvent evaporation and check the design very carefully to ensure the air is not simply the correct volume, but is being introduced and removed efficiently. For rule of thumb calculations, however, you require 1200 cubic feet of air for every pint of solvent. Let us now examine how this applies to a typical machine. The majority of PU machines are likely to operate in the area of 10 yards per minute.

Therefore, with a speed of 10 yards/minute, coating 1 ounce per square yard dry on a web of 60 inches using an 80%/20% solvent/solid solution, we have to evaporate 5 pounds per minute of solvent. This then could require approximately 6000 cubic feet per minute. This quantity of fresh air requires heat as the air will either come from outside the building or from a clean zone within the building.

Obviously if the aid is taken from within the building, this will be replaced with air from outside the building and, therefore, in any calculations an allowance must be made to heat the incoming fresh air requirement to the building.

Let us say it comes from outside the building in the middle of winter. The heat demand for a typical two head plant will be 2.4×10^6 BTU/hour. If the machine were to process vinyl coatings, then the process speed would almost certainly be in the region of 12 to 15 yards/minute, but the heat demand would drop to around 1.0×10^6 BTU/hour because there are no solvents present, hence no fresh air requirements due to maximum re-circulation. After drying, the product is cooled by surface contact with precision machined cooling cylinders.

The manufacture of cooling cylinders varies from stainless steel sheet rolled, welded and finely buffed, to fully machined seamless steel tubes having a wall thickness of 1/2 inch as do many rolled "cans", as they are commonly called. "Cans" are used very successfully in the textile process industry where ovality, eccentricity and like faults are not a serious trouble. All textile materials give or stretch, but when one is handling paper with no give, then one must have accuracy. It is not enough to say paper machines are precision machines so they can process at high speed, they are precision machines because they make paper.

The product must now be either wound up as a sandwich or the laminate separated off the paper, winding paper and product independently. Separation of the product from paper is done in a variety of ways and tends to be a question of plant habit, available space, and what has been learnt on the plant.

We have found that the most successful ingredients for the separation of coated fabrics from release paper compromise the driven separating roller whose speed can be infinitely varied and trimmed in relation to line speed, and effective inspection and access zone to see the material before it reaches the winder.

The winder shafts should be driven by a form of constant tension winder motor so that tension and pressure will be maintained stable throughout the reel. Slipping clutches are used, but these do not wind under controlled tension conditions and also require regular operator attendance. The ideal drive is a DC system using load cells or dancing rolls, but air motors can also be used with effect.

The plant would not be complete of course without a drive and this can either be a shaft system or a multi-motor arrangement. Either system behaves well, and companies produce effectively and consistently regardless of the drive. However, line shafts invariably need guarding which usually means obstructions at floor level. The multi-motor drive on the other hand reduces this guarding almost to zero, keeps the floor clean, but naturally increases installation cost as there are more motors to be wired up.

Control cubicles are usual with plants of this size where all the main control equipment is housed, such as the temperature control equipment and main drive controls. Satellite control points are usually provided adjacent to, say, the coating machines from where the operator can control the machine speed, starting, stopping etc.

Having produced a coated fabric it may be necessary to decorate it. This can be done by surface printing. The ink is applied by a gravure roll to the surface of the urethane coating which is dired, cooled and the product rewound. This operation can be repeated if random designs are chosen, but if a matching design is required than a multi-coloured machine is necessary fitted with print register control. The same comparatively simple machine can be used for lacquering the whole surface to reduce the drag co-efficient or to adjust the colour to a particular shade. Other decorative effects can be provided, for example, embossing and it is here that perhaps there is going to be the greatest variation between urethane fabrics and vinyl fabrics.

Although it is possible to emboss urethane coatings using traditional embossing machinery, it has been shown that surface decoration on urethanes is best produced by tumble or similar violent agitation of the fabric. Some techniques involve immersion of the coated fabric in a water bath with subsequent drying which causes the fabric to shrink, thereby crazing the coated surface. There is, of course, no chance of a repeat design with these techniques, but perhaps the trade is prepared to accept this because the product offers so many other advantages. With regard to the future, there are many developments under way, some at a very advanced stage and it could well be that in 1972 there will be some new methods of producing not just polyurethane coated fabrics, but a wide variety of other coated fabrics as well as new embossing and decorating techniques.

T. J. Brandt

Uniroyal Inc.
Mishawaka, IN

URETHANE-COATED FABRICS FOR SHOE UPPERS

(Received May 1, 1974)

KEY WORDS: Urethane-coated fabrics, shoe uppers, shoe fashions.

INTRODUCTION

The use of urethane-coated fabrics as shoe upper materials has been increasing steadily over the past three years. In the last year and a half, the availability and price of leather goods and PVC has paved the way for urethanes to make a very significant impact on the footwear industry.

In the use of urethane-coated fabrics, the leatherlike qualities are stressed. The soft supple hand, water vapor transmission, surface feel and appearance of some urethanes can rival the finest leathers. In the shoe making process and finished product it is the dissimilar properties that are the concern of this article. Leather is not a product consisting of a thin film coated on a raised napped woven fabric. The only urethane products which have closely approximated the handling characteristics of leather are some of the poromerics on non-wovens which also rival or surpass leather in price.

The basic construction of thin film on napped fabric has given rise to a whole new set of problems. Urethanes as a polymer class are extremely tough and flexible materials. However, most urethane-coated fabrics consist of very thin films; they are delicate and must be handled with due caution.

FABRIC CONSIDERATIONS

In order to maintain the soft hand, urethane films are placed on napped fabrics. A napped fabric has properties that create problems in shoe making. The relatively low and unbalanced stretch of twills and sateens has caused considerable problems of seam popping and pin wrinkling. These fabrics also have a tendency to fray, which is unsightly at best and at worst causes seam weakening. The use of seams that are turned under and double stitched is highly recommended even though this method is inconvenient and more costly. It is felt that this problem can be solved through proper pattern design and cutting techniques. Proper allowances must be made for seaming and sufficient wipe in during lasting and back part forming. The allowances should also be great enough to permit the lasting operations to be carried out using a minimum amount of pressure. When substituting a urethane-coated fabric for either leather or vinyl, experiments should be performed in pattern size, direction of cut, and machine settings.

CATEGORIES OF SHOES

Urethanes are currently being offered to the footwear industry in a wide variety of physical characteristics ranging from soft fashion items to the very tough products which are suitable for men's and children's footwear. For this reason, it is important that footwear manufacturers be aware of these differences and design the total shoe construction with the end use in mind. Most shoes made today could be fitted into four general categories:

1. Fashion Footwear. Shoes that are worn only occasionally or for a short season, such as summertime sandals. The main ingredients for this type of footwear are style and soft hand with sufficient physical properties to perform for a season. Even here, care should be taken not to over-stress the material or have projections that are certain to be abraded.
2. Casual Footwear. Shoes that may be worn daily for part or all of the day. Fashion will again be important in this type of shoe but the emphasis is now shifted to the physical properties of the urethane. Abrasion resistance, adhesion, and flexing resistance will have to be superior to materials used for fashion footwear.
3. Service Shoes. Shoes that will be worn daily for long periods under severe conditions, such as waitress or nurses shoes. Excellent abrasion and flexing resistance and cleanability are important properties that will be demanded of the urethane-coated fabrics. A good urethane system coated at a total weight (skin + adhesive) of 3.5 oz/yd^2 or more may be required for these products.
4. Men's and Children's Shoes. Shoes that will be worn daily under severe conditions and which should have a service life of two years or better. Only superior urethane coating systems should be used with total coating weights in the 5.0–8.0 oz/yd^2 range. Hydrolytic stability, abrasion resistance, adhesion, and flexing will be of prime importance.

The problem of matching coated-fabric materials to specific footwear applications is important because compromises must be made between cost, style and service. There seems to be a general feeling among footwear manufacturers that if a material can be processed with a minimum of problems, a good product will emerge. This may or may not be true. There have been instances where good quality materials have failed under conditions not anticipated. For this reason it is becoming increasingly apparent that a system of classification for urethane-coated fabrics would be desirable. Testing and classification would have to be performed by an independent agency using generally accepted test methods. As the correlation between laboratory testing and in-service performance becomes more sophisticated, a reliable classification system should be quite workable.

CAUSES OF FIELD FAILURES

The basic construction of urethane-coated fabrics of thin film on napped fabric gives rise to two basic problems that are probably the causes for a majority of the field failures. These two problems are adhesion of the film to fabric and pinholes in the urethane surface. These problems are interrelated by the following variable factors: (1) inconsistent nap length and quality, (2) excessive or inadequate laminating pressure, (3) poor formulation of the laminating adhesive, and (4) inadequate drying of the laminating coat which leaves residual solvent that can attack the skin coat. The service failures which have been observed and attributed to these conditions are flex cracking, delamination of the skin from fabric, and snagging or tearing of the skin. Spot checking by the footwear manufacturers can assure the quality of incoming materials. With some technical assistance in sample preparation and minimal expenditure, a simple adhesion testing device can be established. Frequent checking with a magnification device could virtually eliminate the usage of material with excessive pinholes.

Extreme caution is urged in use of solvents in the footwear plant. Most urethane-coated fabrics used today are made from polymers dissolved in solvents. Any excessive contact with solvents in the making operation will damage urethanes. There has been catastrophically expensive damage by (1) solvent wiping the upper for cleaning, (2) spraying solution finishes on the surface, and (3) using excessive quantities of cements in the construction. Obvious damage such as print wipe off is easy to solve by simply being more careful. Not so obvious damage can be caused by entrapment of solvent some place in the shoe construction. Adequate drying time and temperature must be given to any area of the shoe where solvents have been used.

The major coated fabrics suppliers have a responsibility to the footwear trade. First, they must be constantly developing new materials which will upgrade the overall quality of urethane-coated fabrics. Second, they must examine their manufacturing techniques to be sure they are offering the best possible quality. Third, they must familiarize themselves with every phase of the footwear making process and know the requirements. And fourth, they must be capable and willing to offer technical assistance to the shoe manufacturer before and after the sale.

ADVICE FOR FOOTWEAR MANUFACTURERS

The footwear manufacturer can assure himself a more orderly transition into urethane-coated fabrics. He should deal with established manufacturers. Someone who stands behind his product. It's very easy to copy a color or style, but service turns out to be more important.

There should be good communications with the supplier. What do you intend to use the coated fabric for? What's the design? How will you handle it in your plant? These are just a few of the questions. Your supplier can be of most help with the most knowledge of your requirements. Pride, although not invented specifically for footwear manufacturers, can really create bad situations. Knowledgeable people in vinyl or leather techniques are most vulnerable. If urethane-coated fabrics are new materials to your operation, you need help. Your supplier should be your main source.

The use of urethane-coated fabrics in footwear is relatively new in the USA, but it has been used extensively in Europe for several years. The industry's European counterparts suffered all the agonies you may have or expect to experience. An organization called SATRA has been a participant in the growth of urethane-coated fabrics. Although located in England, they have a branch in Toronto. Years of experience are available in required properties, relation of tests to service, and correct shoe engineering. This adds up to correct use of materials. Their services include evaluation of proposed materials, evaluation of sample shoes, and diagnosis of complaints. Your supplier should be a member of SATRA, because there is no way to duplicate this knowledge. Better yet, the shoe manufacturer should be a member. The best of all worlds is the case where both supplier and user are members. In this case, a common language develops through a knowledgeable third party.

Some of the differences between urethane-coated fabrics, vinyl-coated fabrics, and leather have been pointed out. Blind substitution of one for the other is dangerous. Consideration must be given to manufacturing techniques and service requirements. Particularly in the early stages, a footwear manufacturer should have the benefit of technical advice.

L. G. HOLE A. R. PAYNE AND R. J. W. REYNOLDS

Shoe and Allied Trades Research Association, Kettering
Institute of Polymer Technology
University of Loughborough, England

DEGRADATIVE PROBLEMS IN POLYURETHANE FILMS AND COATINGS*

(Received June 12, 1974)

KEY WORDS: Urethane films, urethane coatings, artificial leathers, urethane degradation.

INTRODUCTION

A MAJOR APPLICATION for flexible polyurethanes is in the manufacture of artificial leathers which may be unsupported polymer foam (Porvair), fibre-polymer composites (Xylee, Clarino) and polymer-fabric laminates (polyurethane coated fabrics). These materials have many applications where leather has been the traditional material, e.g. shoes, upholstery, garments, baggage.

There are two main types of artificial leather and they are typified by the poromerics and the polyurethane coated fabrics (PUCF). The former are represented by the cross-section in Figure 1.

This poromeric consists of a non-woven fabric impregnated with a polyurethane microporous foam; this forms the substrate upon which lies a microporous polyurethane foam – this latter about 25–30% of the total thickness of the whole material.

Two types of degradation have been observed in this group: (1) Hydrolysis catalysed by impurities introduced during manufacture which accumulate during wear; (2) Microbiological attack on the polyurethane.

This type of breakdown in the polyurethane foam arising from these factors is shown in Figure 2, and is due to embrittlement of the polyurethane followed by cracking due to flexing of the film.

The structure of the PUCF group is typified by the cross-section shown in Figure 3.

As a rule the structure consists of a one-component polyurethane which forms the 0.05 mm thick surface film. This film is attached to the fibrous pile of the substrate by a polyurethane adhesive film; this latter is commonly of the two

*Paper presented at the European Plastics and Rubber Conference, Paris, France, June 4–12, 1974.

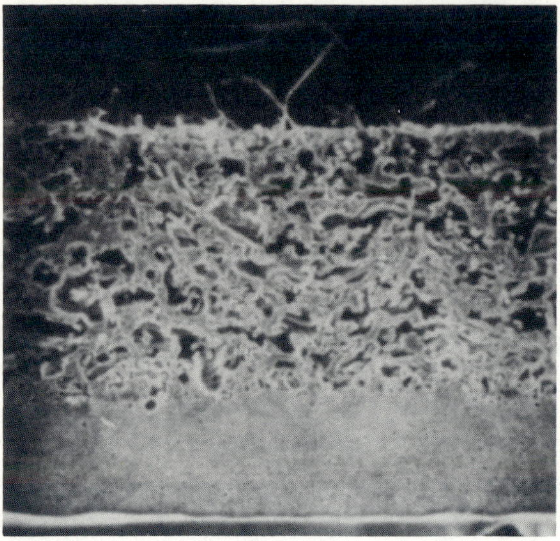

Figure 1. Cross-section of a typical poromeric artificial leather.

Figure 2. Cross-section of the degraded microporous foam.

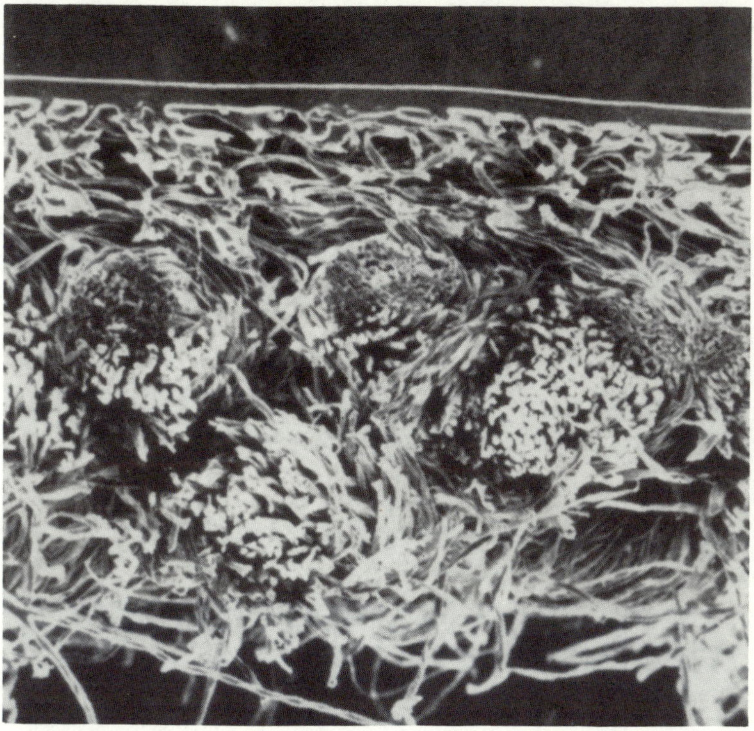

Figure 3. Cross-section of a typical PUCF.

component type. One component polyurethanes are usually significantly more resistant to hydrolysis than the two-component types, thus failure of the adhesive tie-coat in these materials, due to hydrolysis, is more often observed than hydrolysis deterioration of the film.

These films are susceptible to deterioration by UV light, however, and our experiments on this aspect are described later.

THE HYDROLYSIS OF POLYURETHANES

Myagkov and Pakshver [1] showed that the dilute acid hydrolysis of Polycaprolactam was dependent upon [LACTAM] $[H_3^+O]$ but found that $[H_3^+O]$ did not change when excess of acid was used. Thus lactam hydrolysis was postulated by them to take place by electrophilic attack on the nitrogen atom:

$$\underset{\vdots \atop H_3^+O}{R'.CO.NH.R} + H_2O \longrightarrow R'\,CO.OH + R.NH_2 + H_3^+O \quad \ldots\ldots\ldots\ldots\ldots\ldots \text{(i)}$$

In the case of polyurethanes based upon an ester precursor, two linkages are susceptible to hydrolysis – the urethane and the ester links. Roussel [2] has shown that ester links are more easily hydrolysed than urethane links, however, the presence of amine breakdown products when some polyurethanes are hydrolysed at 120°C indicates the scission of some urethane links.

It may be postulated that the polyurethane may undergo nucleophilic attack if the nitrogen in the urethane link is already involved in an inter or intra-chain hydrogen bond as indicated by Equation (2).

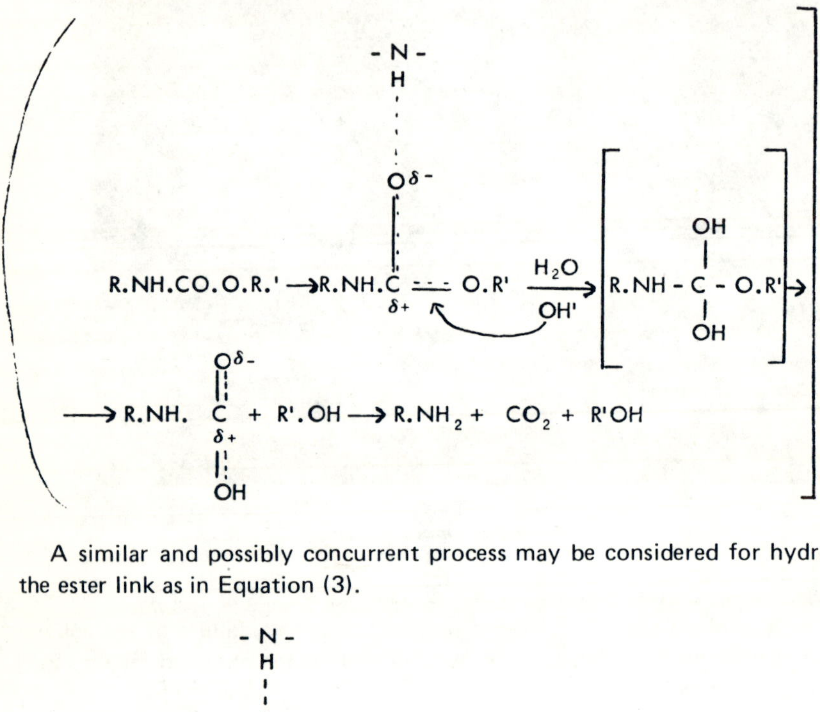

.... (2)

A similar and possibly concurrent process may be considered for hydrolysis ot the ester link as in Equation (3).

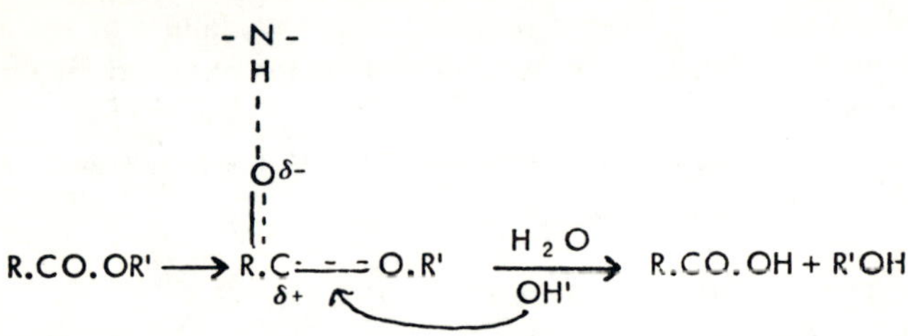

.... (3)

Thus catalysts for the nucleophilic reaction will clearly accelerate the degradation rate and it has been shown by Hole et al, [3] that various substances in human sweat, e.g. urea, and in sebum, e.g. oleic acid, catalyse the hydrolysis of polyester urethanes. It was also shown that time-temperature superposition methods could be used to investigate hydrolysis rates and the activation energies for polyurethane scission where changes in tensile energy were taken to represent k (the velocity coefficient of the hydrolysis reaction). Similar methods were used to investigate the effect of metal ions as hydrolysis reaction catalysts.

The use of ethylene diamine tetracetic acid (EDTA) was proposed [4] as an anti-hydrolysis stabiliser and its effectiveness is demonstrated in Figure 4.

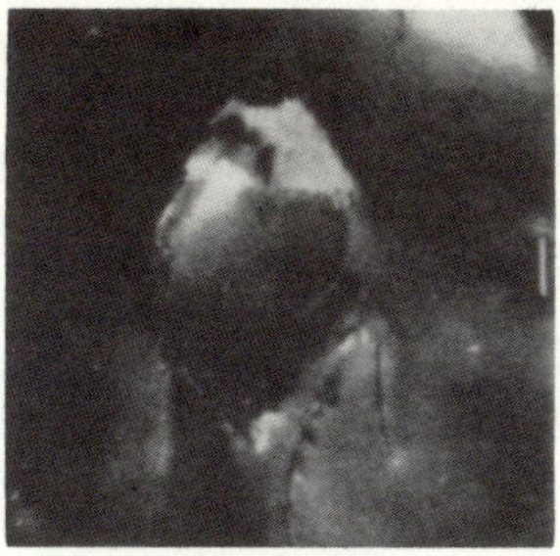

No Stabiliser
100 flexes

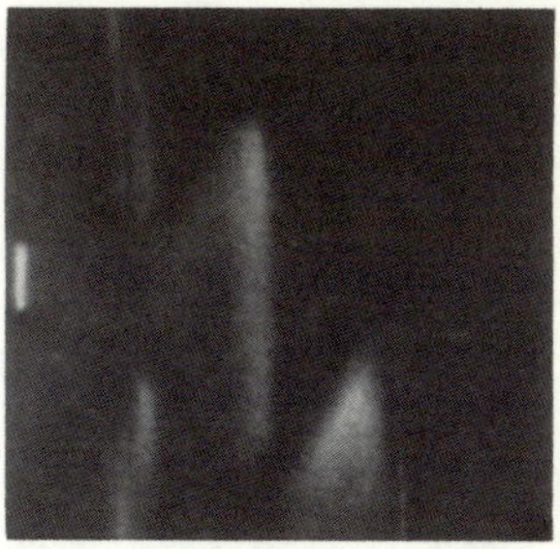

With Stabiliser
1 million flexes

Figure 4. Influence of $EDTANa_2$ on the Hydrolysis of a urethane in the presence of $Fe^{++}/^{+++}$ion.

The control samples were hydrolysed in urea solution (1.2 gpl) at 120°C; the test sample was similarly degraded, but with 2% W/V $EDTANa_2$ added to the urea solution. The chelation of the iron ions has clearly provided significant protection to the polymer film. This effect is further demonstrated in the polymer hydrolysis curves in Figure 5.

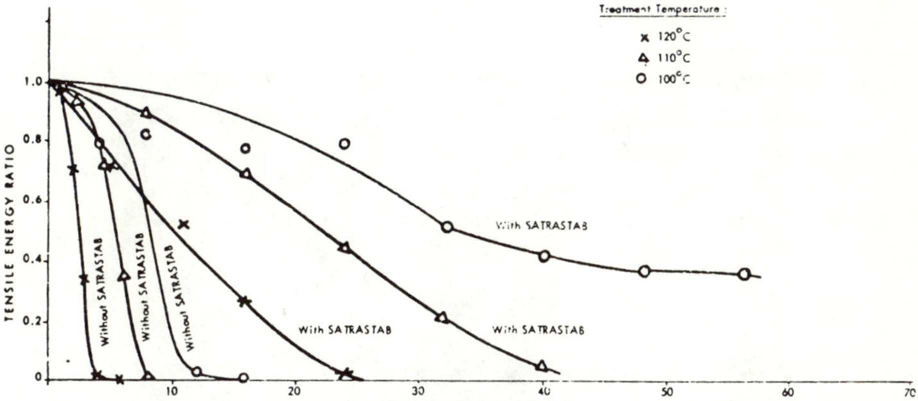

Figure 5. The influence of SATRASTAB on the stability of a commercial poromeric when degraded in urea solution (1.2 g/l).

ESTERS OF EDTA AS STABILISERS AGAINST HYDROLYSIS

Microcrystalline $EDTANa_2$ has some disadvantages for use in polyurethane films: (1) $EDTANa_2$ is not miscible with polyurethane systems, and the microcrystalline form lowers the mechanical strength of the film, (2) $EDTANa_2$ is leached out of the polymer by water at a relatively rapid rate, (3) $EDTANa_2$ cannot be used with metal catalysed one-component urethane reactions.

It was considered that fully or partially substituted esters of EDTA would overcome these disadvantages. The sequestering power of the EDTA should not be interfered with since the nitrogen and oxygen atoms remain free for chelation and are not sterically hindered by the proposed alkyl groups as indicated in Figures 6 and 7.

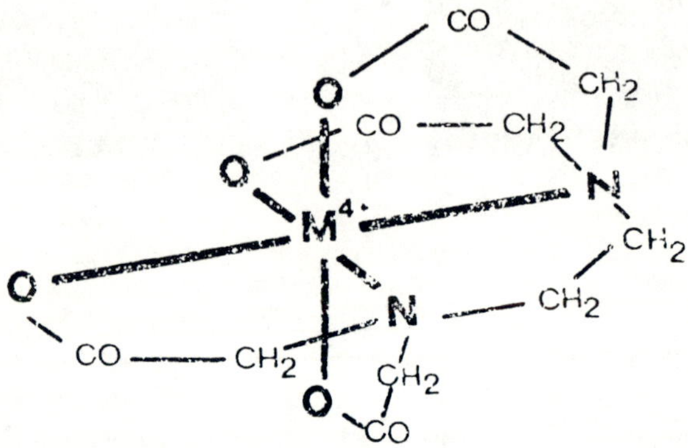

Figure 6. Structure of EDTA ester in metal chelate form.

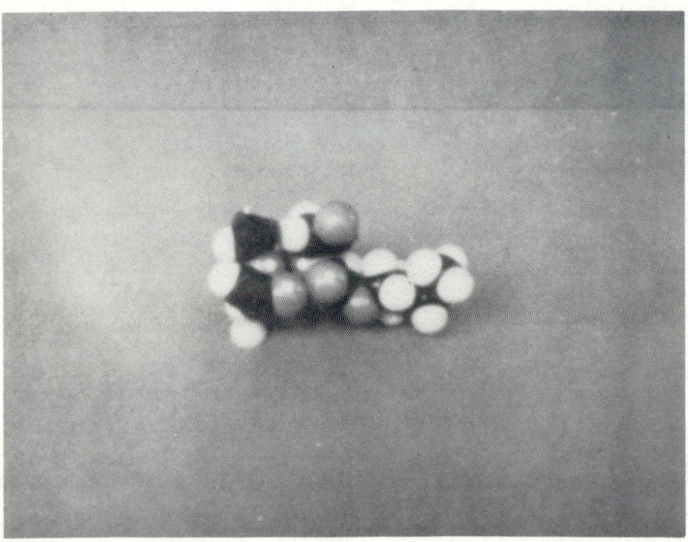

Figure 7. Molecular model of alkyl ester of EDTA.

Under the conditions of the practical application of these esters as stabilisers, however, it is to be expected that their negligible solubility in water would prevent the formation of the chelate, thus it may be possible to employ them in one-component systems, i.e. added to the reactants prior to urethane formation. The tetra substituted methyl, isopropyl and n-butyl were prepared according to the method of Bersworth [5], the method was found to be convenient and provided fairly good yields.

These EDTA esters will hydrolyse under those conditions which hydrolyse the polyurethane and the aim was to incorporate an ester in the polymer such that the ester hydrolysed more rapidly than the polymer. Thus hydrolysis of the ester would produce free EDTA in a highly dispersed form within the polymer – this would achieve two objectives: (1) chelate any metal ion catalysts present, and (2) provide some control of pH in the 5–7 region to inhibit any basic contaminants around or within the polyurethane.

The hydrolysis rates of the tetra methyl and tetra n-butyl esters are illustrated in Figure 8. This shows the different rates of the two esters under extremes of conditions and demonstrates the anticipated higher stability of the n-butyl ester.

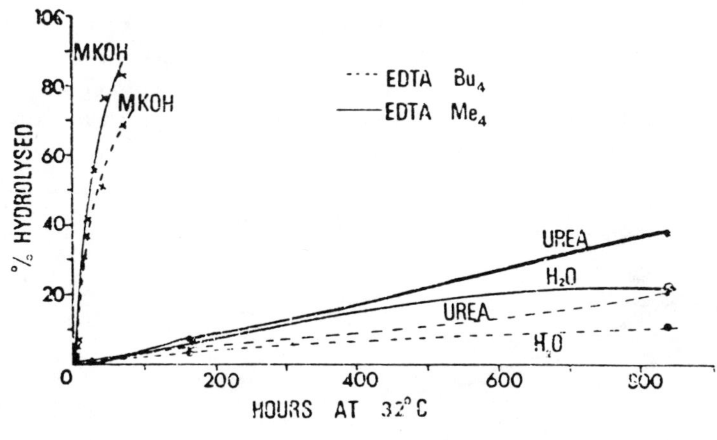

Figure 8. Rates of hydrolysis curves for the tetra Me_4 and nBu_4 esters of EDTA.

The gradual conversion of the EDTA esters to free acid can be expected to cause loss of stabiliser by leaching out due to the relative solubility of the free acid in water. Experiments showed that after 12 hours immersion in distilled water significant loss of EDTA occurred as follows:–

Microcrystalline $EDTANa_2$	:	65% removed from polymer by leaching
Tetra methylester of EDTA	:	15% removed from polymer by leaching
Tetra n-butyl ester of EDTA	:	11% removed from polymer by leaching

The esters are, therefore, much more persistent in the polyurethane than the disodium salt of EDTA, and have the advantage of producing the free acid in just those regions where hydrolysis is likely to occur, moreover in a state of extreme dispersion.

Tests on a Polyester Polyurethane

The effectiveness of the methyl and n-butyl tetra esters was investigated using a polyurethane based on a polyester prepolymer viz: 1.4 butane diol (95%) 1.2 propane diol (5%) adipate, this was reacted with 4.4 methylene diphenyl diisocyanate (MDI). The resulting isocyanate terminated polyester was chain extended with 1.4 butane diol. The polyurethane had the following properties:

Number average molecular weight	:	46 200
Tensile strength	:	430 kg/cm^2
Elongation at break	:	1 250%

This polymer in solution was treated with the catalyst additives viz: Sn^{++} as stannous octoate and a brown pigment containing ferric oxide, and also with the appropriate EDTA compound, prior to coating onto glass and drying. Polymer films of about 0.2 mm thickness were thus prepared for the ageing tests.

In a typical hydrolysis experiment a film containing 100 ppm Sn^{++} and 2% EDTA (in the form of the tetra methyl or tetra butyl ester) was hydrolysed at 120°C in steam for up to 15 hours. The fall in tensile strength with exposure time is shown in Figure 9. The results of a similar experiment in which hydrolysis was carried out in urea solution (1.2 gpl) is shown in Fiqure 10.

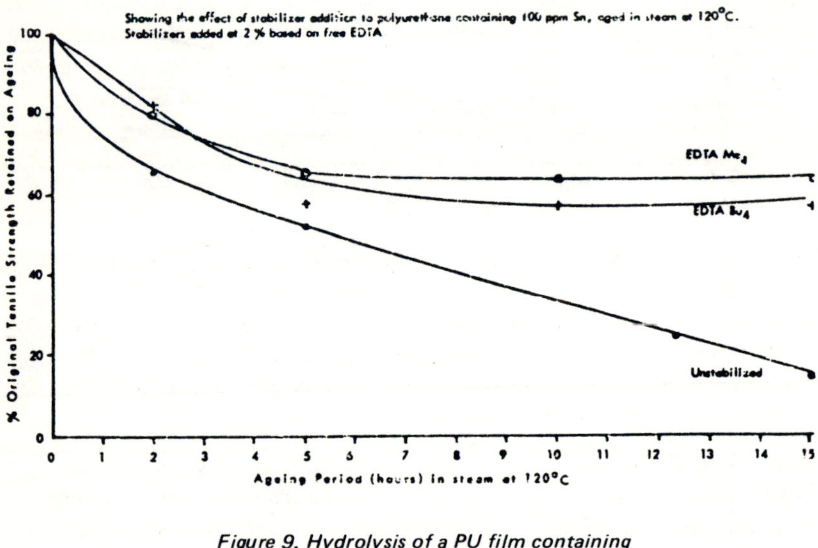

Figure 9. Hydrolysis of a PU film containing Sn^{++} and EDTA esters. Steam 120°C.

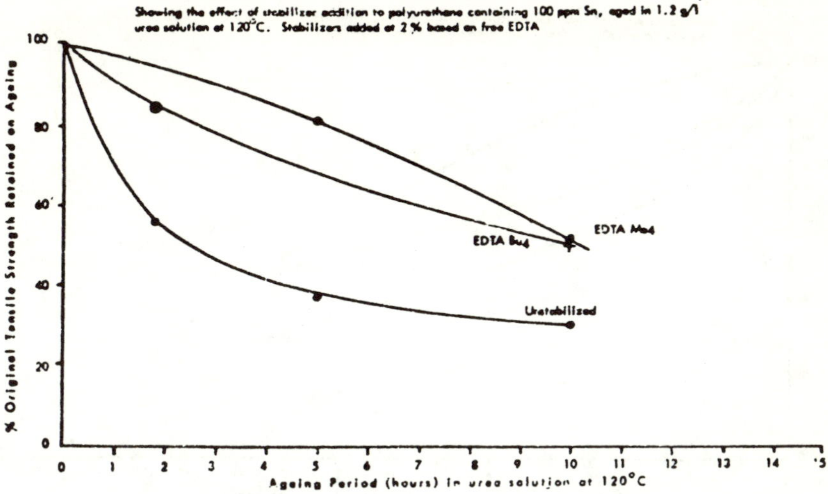

Figure 10. Hydrolysis of a PU film containing Sn^{++} and EDTA esters. Urea solution (1.2 gpl) 120° C.

The results in Figures 9 and 10 show the increases in the time for the polymer to reach a particular level of strength reduction (see Table 1*).

The figures in Table 1 demonstrate a significant improvement in stability of the polyurethane – an improvement which is more pronounced in the presence of two catalysts – Sn^{++} and urea. The reason for this latter may be associated with loss of degradation products (autocatalysts) from the polymer degraded, whilst immersed in an aqueous solution. In steam the degradation products will accumulate in the polymer and if of a catalytic nature will increase the rate of degradation. A further factor may be that the EDTA esters hydrolyse more rapidly in the presence of the urea due to the ammonia produced under the hydrolysis conditions, thus releasing more EDTA for chelation of the Sn^{++}. This postulate is supported by the greater effect of the tetra methyl ester, which was previously shown (Figure 8) to hydrolyse more rapidly than the tetra n-butyl ester – hydrolysis rate is a function of the alkyl chain length, tetra methyl being the least stable. In practice it may be advantageous to employ mixtures of the C_1 – C_4 alkyl tetra substituted esters; furthermore, the use of partially substituted esters of the type may be of value.

The results of further experiments in which an Fe^{+++} containing brown pigment was used as the catalyst in place of Sn^{++} are shown in Figure 11. In this case the tetra methyl, isopropyl and n-butyl esters were examined.

These curves provide further evidence that these EDTA esters behave as hydrolysis stabilisers and that the effectiveness increases as the number of carbon atoms in the alkyl substituent. This is shown in Table 2.

*All Tables at end of article.

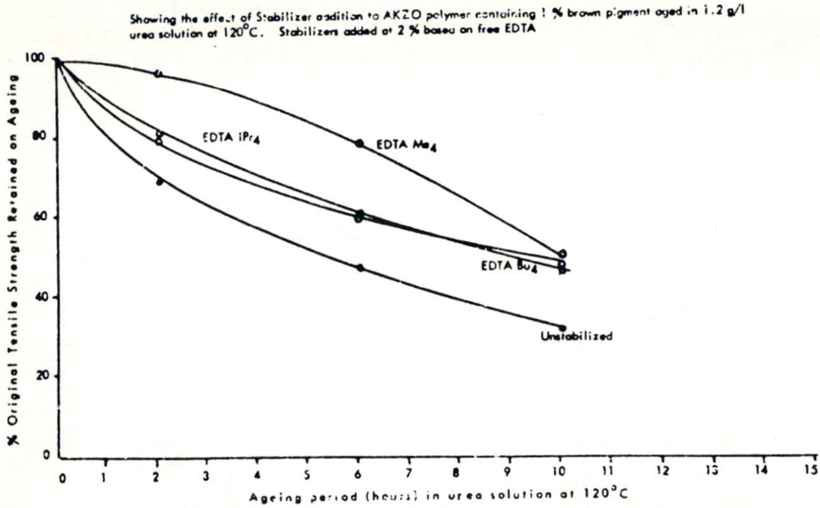

Figure 11. Hydrolysis of commercial poromeric PU film containing brown pigment (Fe^{+++}) and EDTA esters. Urea solution (1.2 gpl) 120° C.

THE PHOTODEGRADATION OF POLYURETHANE FILMS

The one-component polyurethane films used on PUCFs are in practice relatively resistant to hydrolytic degradation of the type discussed above, however, the two-component tie-coat adhesives do give problems due to hydrolysis, thus the previous discussion is relevant to these adhesive polymers. Whilst UV degradation (UV discolouration is not considered here) is apparently not a problem with the microporous polyurethane systems used in the poromeric artificial leathers, it has caused difficulties with the one-component polyurethane films used in PUCFs. Many failures observed in upholstery PUCFs have been attributed to the effect of UV absorption by the polymer.

The exposure of PU to UV light leads to visible effects such as discolouration and/or surface cracking and a general reduction in mechanical properties. In the absence of oxygen photolysis reactions are possible depending upon the wavelength of the light; however, in the presence of oxygen photo-oxidation reaction predominate and occur over a wider range of wavelength. Thus degradation will occur by chain scission and crosslinking both ultimately leading to polymer embrittlement.

The bond energies in a polyurethane are such that UV of the appropriate wavelength will produce catastrophic degradation; however, in practice poor absorption and low quantum yield in the polymer provides for a significant level of stability.

In practice UV absorption must be associated with a photo-oxidation degrading mechanism, although photolysis of the polymer molecules may be the initiation stage of the photo-oxidation reaction. Trozzolo [6] has suggested that the following may occur: (1) Primary photolysis – direct absorption of photons by the polymer molecule – in the normal state or an oxygen-polymer or ozone-polymer complex,

$$\text{e.g. } RH + O_2 \xrightarrow{hV} [RH \pm O_2] \longrightarrow \text{reactive intermediates} \quad (4)$$

Also, secondary photolysis p impurities, e.g. transition metal ions, in the polymer may absorb photons and thus be oxidised and then cause scission of the polymer chain. If the oxidised impurity is a better UV absorber the system becomes autocatalytic. Schollenberger and Dinlorgs [7] have suggested a quinone-imide route for the photolysis of a polyester urethane and the development of colour in the polymer provides some support as well as infra-red and UV spectroscopy which indicate initial loss of aromaticity with formation of the coloured quinonoid structure (see Equation 5).

The quinone-imide may form crosslinks with urethane groups in adjacent unexcited chains (Equation 6).

Various expedients are employed to prevent photodegradation, such as: UV absorbers and quenchers (deactivate excited states in the polymer) Light screens, e.g. surface coatings, pigments.

In the commercial materials mixtures of antioxidants and UV absorbers are incorporated in the polymer. Pigments are usually included and carbon black is known to be particularly effective as a screen.

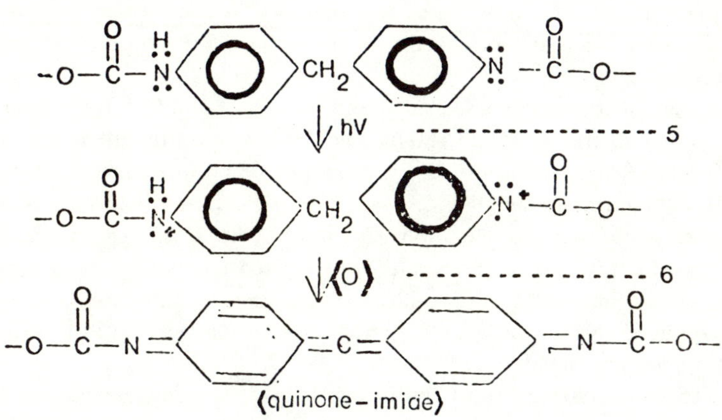

....(5) & (6)

Some Results of the Photodegradation of a Polyurethane

A commercial polymer (Crestopol 112 Scott-Bader Ltd.) was used and appropriate polymer, thus films were made using a Werner-Mathis Coating Machine. In the pigmented films a brown microlith pigment (45% dispersion in MEK) was added to the PU solutions to give 10% pigment solids on dry polymer weight. Photodegradation was carried out in the Xenotest 150 machine in dry and moist

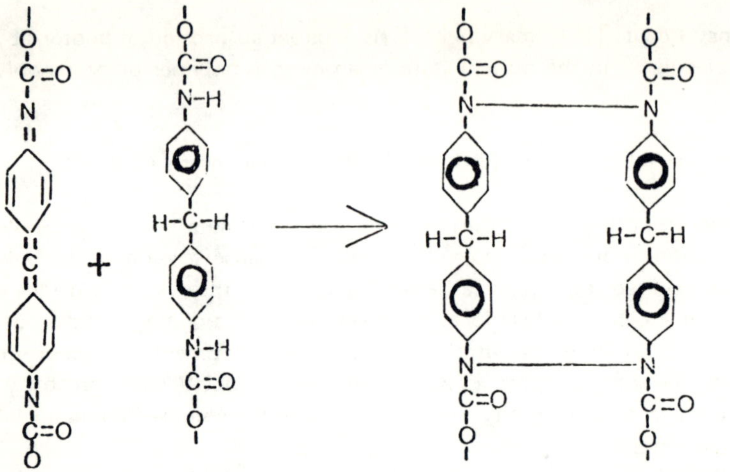

... (7)

conditions. Changes in the polymer were observed by measurement of the tensile properties of the exposed polymer.

Initially the UV absorption of four polymer films of various thicknesses (0.036 mm–0.24 mm) were measured using an Oxalic acid-Uranyl sulphate actinometer [8] and the results demonstrated that the Beer-Lambert Law was obeyed at a film thickness of up to about 0.2 mm at 60 mins exposure. Thicker film showed a trend for a negative deviation from the Beer-Lambert Law.

Changes in the mechanical properties of the exposed films were determined in the pigmented and clear state and one result for the latter in wet and dry exposure is shown in Figures 12 and 13.

These results demonstrate the adverse effect of high humidity (including a period of "rain" cycle in the Xenotest 150 machine) on tensile strength and breaking elongation – particularly the latter. This is of practical significance for materials which have clear polyurethane lacquer finishes e.g. PUCFs, patent leathers, where surface strains over a fold such as occurs in wear can be up to 20% linear extension. The cyclic folding which occurs in shoes coupled with the combined UV-moisture degrading effect which Figures 12 and 13 illustrate could lead to flexural failure of the PU film and indeed this is not an uncommon type of damage found in PUCF materials in shoes and upholstery.

The addition to the polyurethane of the brown Microlick pigment is shown by Figures 14 and 15 to produce an improvement in stability as measured by elongation properties and the effect is highly significant at high humidity under which conditions the clear film is degraded relatively quickly. At the low humidity the unpigmented film has lost 50% of its elongation after about 480 hours exposure in the Xenotest 150 whilst at the high humidity the 50% reduction was reached after about 260 hours. These results indicate an association between moisture and UV absorption – the screening to UV provided by the pigment clearly minimises the

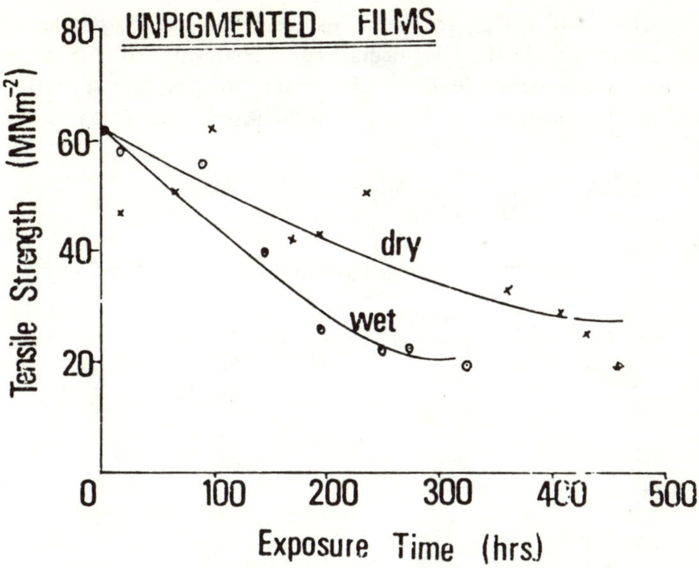

Figure 12. Photodegradation of clear PU film at 50–55% rh/45° C and 98% rh/28–45° C (Change in tensile strength).

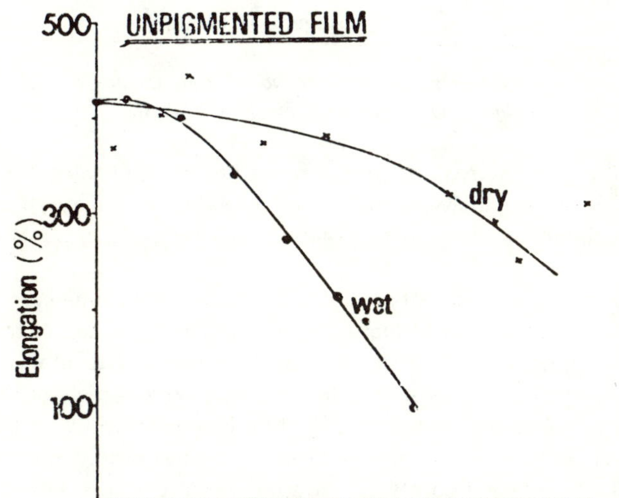

Figure 13. Photodegradation of clear PU film at 50–55% rh/45° C and 98% rh/28–45° C (Change in breaking extension).

effect of moisture and the absence of moisture minimises the degrading effect of the UV light. It is feasible that the degrading mechanism results from the creation of (H_3O) by UV absorption which might scission the polymer at urethane links in a way similar to that proposed by Myagkov and Pakshver (see Equation 1).

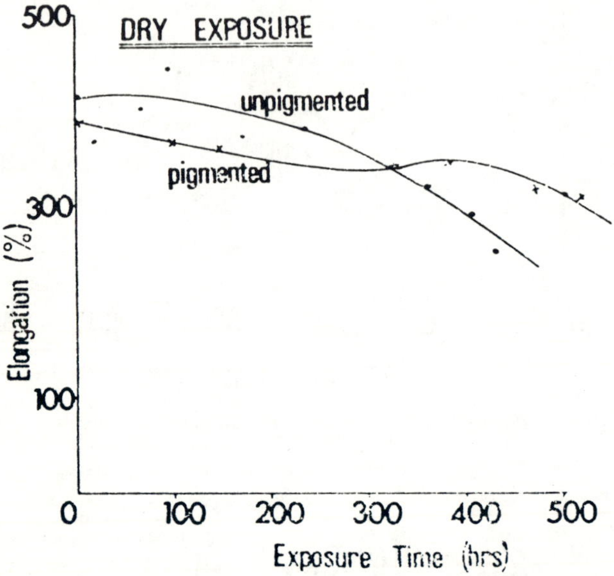

Figure 14. Pigmented and unpigmented film. Change in breaking elongation at 50–55% rh/45° C.

The degradation in these films was also marked by yellowing (in the unpigmented film) and a progressive reduction in solubility in dimethylformamide — evidence of crosslinking reactions taking place during exposure to UV (see Equation 7).

Similar experiments on pigmented films obtained from commercial PUCFs have demonstrated the much greater reduction in elongation properties of the film after a given exposure, compared with the reduction in tensile strength. This is practically useful information because of the folding/flexing which occurs in most artificial leather applications. Reductions on film extension will very likely reduce the flex cracking performance of the material — this has been demonstrated with poromerics which have been hydrolysed to various levels and then flexed.

CONCLUSIONS

The main points are as follows: (1) Some polyurethanes can be protected against hydrolysis by incorporating the tetra methyl, tetra isopropyl and tetra n butyl

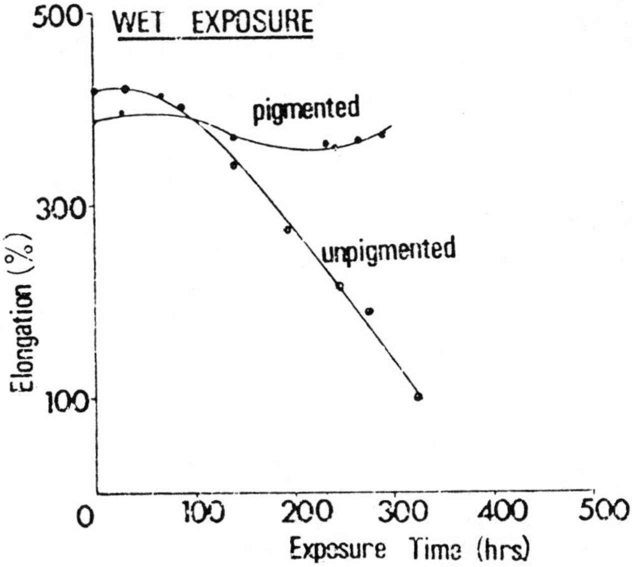

Figure 15. As Figure 14 but changes occurring at 98% rh/28–45° C.

esters of EDTA into the polymer, (2) the efficacy of the EDTA esters reduces as the number of carbon atoms in the alkyl substituent is increased, (3) the rate of loss of the EDTA esters from the polymer is slow even when the film is immersed in water, (4) in the photodegradation experiments it has been found that thin clear polyurethane films obey the Beer-Lambert Law. The films show mechanical deterioration, especially in elongation properties, as a result of UV exposure and the presence of moisture increases the rate of degradation considerably, (5) the addition of a pigment to the polyurethane causes a pronounced improvement in stability in dry and wet conditions although the rate of deterioration is greater in the presence of moisture, (6) some of the experimental results enable speculation that improved UV resistance would be obtained by incorporating a suitable proton acceptor in the pigmented polymer.

Table 1. Increase in time for polymer to degrade to 65% of original tensile strength.

	Steam 120°C	Urea solution 120°C
Polymer + Sn^{++}	2 hours (x1)	1.25 hours (x1)
Polymer + Sn^{++} + EDTA Bu_4	4.3 hours (x 2.1)	5.75 hours (x 4.6)
Polymer + Sn^{++} + $EDTA_2$ Me_4	5.5 hours (x 2.7)	6.80 hours (x 5.4)

Table 2. Increase in time for a commercial polymer to degrade to 65% of original tensile strength

	Urea solution (1.2 gpl) 120°C
Polymer + pigment	2.5 hours (x 1)
Polymer + pigment + EDTA BU_4	4.3 hours (x 1.7)
Polymer + pigment + EDTA Pr_4	4.8 hours (x 1.9)
Polymer + pigment + EDTA Me_4	7.9 hours (x 3.2)

REFERENCES

1. V. A. Myagkov and A. B. Pakshver, Kolloidn Zh 19752 *14* 172.
2. P. A. Roussel, "Ageing of Polyester Urethane Foams," Bulletin Hr-15 Du Pont Co. (Elastomers Division).
3. L. G. Hole, G. Kann, P. J. F. Dawkins, and R. J. W. Reynolds, International Rubber Conference F 3 1-6 Brighton, 1972.
4. L. G. Hole, Brit. Patent.
5. F. C. Bersworth, U.S. Patent No. 2428 353, 1943.
6. A. M. Trozzolo, "Stabilisation against oxidative photodegradation," Polymer stabilisation Interscience, 1972.
7. C. S. Schollenberger, K. Dinbergs, SPE Transactions, Du Pont, 1 January 1961.
8. W. G. Leighton and G. S. Forbes, J Phy. Chem. 1938 *42* 701.

FRANK J. GASPARRINI

Millmaster Onyx Corporation
RO Polyurethane Specialites Div.
624 Schuyler Avenue
Lyndhurst, New Jersey

THE DESIGN OF SOLUTION POLYURETHANES FOR TEXTILE COATINGS

(Received September 16, 1974)

ABSTRACT: The basic parameters and physical properties of various solution urethane polymers for textile applications are discussed. This article also classifies different urethane polymers and demonstrate the performance characteristics of each. The importance of applications data, performance testing, and the ability of the urethane chemist to work directly with the textile coater are also examined. New areas of interest, such as urethane topcoats for vinyl and water-based urethane lattices, are discussed.

KEY WORDS: Textile coatings, urethane coatings, apparel, footwear.

INTRODUCTION

THIS PAPER REVIEWS the design of urethane polymers to meet the requirements of the various markets for coated fabric. In the sequence of supply from the manufacturers of raw materials, the coaters, the fabricators of finished products to the merchandisers, a combination of requirements arises. The buyers of the final products are motivated by aesthetics, performance and price. They look for the colors, surface effects, hand and softness of the articles which they purchase, whether because of the dictate of current fashion, or due to their own preferences. Colors change with the season of the year. Surfaces range from patent to matte, and from glove softness to the sturdier, stiffer products designed for high performance use such as men's shoes. Ladies buying fashion shoes will choose a line of colors to match their different outfits. Colors and styling are their main objectives. These shoes will be worn only occasionally, for short periods and normally under conditions not involving extreme flex or exposure to scuffing. Thus, performance requirements in this case are relatively moderate. On the other hand, a nurse buying uniform shoes will wear these day after day for full periods of duty covering miles of walking during those periods. She will demand that these shoes be comfortable and will expect at least many months of wear without marring of the surface or discoloration and certainly no delamination or cracking. Some will be worn in cold climates and must perform well under these conditions. Shoes must slip readily if the heels accidentally rub together during walking. Failure to do so has caused serious accidents to the wearer.

PERFORMANCE REQUIREMENTS

Coated fabric for upholstery adds the requirement of use for long periods, often in a warm, humid climate. In apparel, for example coats, the buyers purchase on the basis of hand, drape, color, surface grain and, of course, tailoring. They will expect long wear without scuffing or picking, and no delamination or cracking even in cold weather. They, in many cases, will have the coats dry cleaned and will expect them to return from the cleaners unchanged in appearance or hand.

Thus, from the top of the scale, a set of requirements start to generate which become the objectives of polymer design. In most cases, the polymers must be highly resistant to exposure to sunlight. They must be sufficiently soft to provide good hand and drape, yet must be tough and strong enough to have good flex life and resistance to scuffing and abrasion. The various layers of urethane applied to the fabric must adhere well to each other under conditions of repeated flexing even when wet or very cold. The surface of the polymer films must be capable of taking on a glossy, patent effect, yet must slip well when rubbed together. In apparel, the polymers as a coating on fabric, must be resistant to dry cleaning solvents such as perchlorethylene. High levels of hydrolytic stability must be incorporated in these polymers such that they are able to resist both foot perspiration in shoes and long term exposure to high humidity in upholstery.

Additional performance requirements arise during the fabrication of the coated fabric into finished articles. Generally the manufacturer of footwear is the most demanding in this respect. During shoemaking, the coated fabric is pulled under high tension to conform to the various curves of the shoe surface and must not change in appearance under these conditions. Toe posting must not impart any permanent indentation to the coated cloth. Various compositions, some containing potent solvents such as methyl ethyl ketone and tetrahydrofuran are applied to the under surface of the fabric to form hard box toes and heels. Excess sole attaching adhesives are wiped from the urethane surface with rags wet with methyl ethyl ketone.

Proceeding one step backward, the coater has requirements specific to his operation. These include the nature of the solvents which his ovens can handle as well as the type dictated by EPA and OSHA regulations. His coating heads whether knife, gravure or reverse roll dictate the solids-viscosity ratio which he can tolerate.

Thus, from the many operations through which coated fabric must pass, many specifications arise. For the purpose of this article, classification by end use is appropriate. The high performance requirements for footwear and upholstery are very similar and are summarized as follows: flex life, abrasion resistance, hydrolytic stability, solvent resistance, light stability.

Likewise, the lower performance markets of apparel and handbags have somewhat similar sets of demands. The key needs for these applications are listed in Table 1.

Table 1. Desirable Performance Properties in Urethane Content Apparel

Flexibility	Light Stability
Hand and Drape	High Slip
Dry Cleanability	Low Tack

The performance needs discussed to this point are related to articles made from transfer coated fabric. Direct coated fabric is converted into quite different finished products with additional requirements arising from their use. All must be waterproof. For consumer products such as rainwear and skiwear, permeability, hand, drape and styling are key features. Abrasion resistance and tear resistance are particularly critical in industrial products such as rainwear and tarpaulins. Air holding properties are essential in inflatable goods such as life vests and airplane escape chutes. This leads to another set of requirements (Table 2).

Table 2. Desirable Performance Properties in Urethane Direct Coated Fabrics

Flex Life	Air Holding Properties
Abrasion Resistance	Heat Sealability
Water Resistance	Hand and Drape

This paper discusses how, as far as possible, the solution to this complex puzzle is reduced to a science. The raw materials which are commercially available to the urethane chemist will be reviewed. The variety of tensile modulus curves which can be developed through such raw material combinations will be demonstrated. The relation of tensile-modulus to end use of the urethane whether as skin coat, adhesive coat or top coat will be shown. Tests values such as flex life and abrasion resistance obtainable with combinations of these coatings on fabric will be tabulated. Concentration will be on footwear systems because of the rigid, complex demands of this market. Effect of chemical composition on such basic properties as hydrolytic stability and solvent resistance will also be covered. Systems have been chosen which have been proven by commercial use and which thus have undergone the tortures of shoe making and the ultimate test of wear by the consumer.

The paper will then cover polymer design, the first link in the chain of this long process to provide saleable, high performance products. In the next link, the coater must combine these polymers properly with a suitable fabric or the best efforts of the resin producer can be negated. The chain can be broken at the next link for example, the shoemaker, if he constructs products of poor quality or styling. Finally none of the links can exist without the merchandizing efforts of the companies selling the finished shoes, coats and furniture.

With this chain of dependence and requirements there must be a high degree of cooperation and communication. Coaters must work very closely with the polymer manufacturers. They must outline in detail their specifications and test methods. Samples of the exact fabric which they are using are essential. Composites are prepared by both parties and performance results are compared. Evaluations are continued until the proper balance of properties is achieved. Coaters report to the resin producer on the performance of the coated fabric during shoemaking and later the results of wear trials. The polymer manufacturer must depend on the coater for contact with these other links in the chain. Thus they obtain information on changes in styling and specifications originating with the consumer, merchandizer and even the coaters themselves.

Many coaters permit the resin producer to be present at line trials to observe the performance of their systems during the coating operation. Success in providing high performance polymers to coaters usually is in direct proportion to the cooperation between these two links. It cannot be emphasized strongly enough how important such a relationship is to success. Similar close contacts which exist further along the line between coater, shoemaker and merchandiser are also essential.

Formulating Urethane Coatings

The two basic reactions involved in the synthesis of polyurethane polymers are the reaction of an isocyanate with a hydroxyl bearing glycol to yield a polyurethane. The other is the reaction of an isocyanate with a material containing an amino end group to yield a urea. These two reactions are illustrated below.

$$\mathrm{RNCO + HO - R^1 \longrightarrow R\overset{\overset{H}{|}}{N} - \overset{\overset{O}{\|}}{C} - O - R^1} \qquad \text{URETHANE}$$

$$\mathrm{RNCO + H_2N - R^1 \longrightarrow R\overset{\overset{H}{|}}{N} - \overset{\overset{O}{\|}}{C} - \overset{\overset{H}{\|}}{N} - R^1} \qquad \text{UREA}$$

There are other reactions which can take place such as the formation of biuret and allophonate linkages which are generally undesirable.

The three general classifications of raw materials which are used in the polymerization of polyurethanes are polyols, diisocyanates and chain extenders. Each of these has a direct bearing on the properties of the resultant polyurethanes; however, the polyol and the diisocyanate can be considered to form the backbone of the polymer chain. The chain extender is responsible for the final molecular weight of the polymer which is necessary for ultimate performance properties.

The classifications of polyols can be considered to include both polyethers and polyesters. Since polycaprolactones are basically polyesters, they are included in that discussion. Overall, polyesters impart the best combination of properties for urethanes for textile coatings and in general, are chosen for their resistance to solvents and outdoor weathering, including oxidation. They also provide the urethane chemist with versatility in design of molecular structure. By varying the chemical composition of the ester, he can formulate polymers which will range from hard and brittle to very soft and extensible.

On the other hand, polyethers impart slightly better low temperature properties to the polymer. In addition, polyethers have an edge on polyesters with regard to hydrolytic stability. However, by choosing a polyester of the proper chemical composition, polyurethanes can be designed which will pass severe humid aging tests. In essence, this can be accomplished in several ways, such as lengthening the carbon chain between ester linkages and/or by using glycols containing secondary hydroxyls. As a result, polyesters are predominantly used in the formulation of solution polyurethanes for textile coatings. The different polyurethane properties obtainable with these two types of polyols are summarized in Table 3.

Table 3. Comparative Polyurethane Properties vs Polyol Type

Polyesters	Polyethers
Solvent Resistance	Hydrolytic Stability
Resistance to Outdoor Weathering	Low Temperature Properties
Variety of Polymer Properties	

For reasons of cost, most polyesters are formulated around adipic and phthalic acids and combinations of the two. Dibasic acids such as azelaic or sebaic can also be used to impart other special properties but generally have been too costly.

The straight chain glycols such as ethylene glycol, 1,4 butanediol and 1,6 hexanediol, tend to give urethane polymers which are somewhat hard and crystalline with relatively high tensile strengths. In addition, as indicated above, the longer the glycol chain, the better the hydrolytic stability. On the other hand, solvent resistance is best with a short chain glycol. Glycols with secondary hydroxyls also improve hydrolytic stability and those with pendant alkyl groups give urethanes that are softer and more flexible. The use of diethylene glycol with its ether linkage is an additional method of obtaining softness and flexibility.

Besides affecting the performance properties of the finished urethane, the structure of the glycol will also affect the physical state of the polyester. Thus, a straight chain glycol adipate such as ethylene glycol adipate, produces a relatively high melting, waxy polyester. However, a propylene glycol adipate with its pendant alkyl group is a liquid polyester, as is the diethylene glycol adipate because of its ether linkage. It is also possible to manufacture a liquid polyester which will give high tensile urethanes by the use of eutetic mixtures of raw materials. The effects of glycol structure are summarized in Table 4.

Table 4. Effects of Glycol Raw Materials

Glycol Type	Physical Form of Polyesters	Polyurethane Properties
Straight Chain	Wax	Hardness High Tensile Strength Crystallinity
Pendant Alkyl Groups		Softness
	Liquid	Flexibility
Ether Oxygen Containing		Low Temperature Properties
Mixed Glycols	Soft Solid to Liquid	Moderate Softness High Tensile Strength Flexibility

As indicated in Table 4, both the ester and the isocyanate used have a substantial effect on the final properties of the urethane. Isocyanates can be generally divided into two main types, aromatic and aliphatic or cycloaliphatic (non-yellowing). The aromatic isocyanates commonly used are TDI (toluene diisocyanate) and MDI (diphenylmethane diisocyanate). The non-yellowing types of isocyanates most commonly used are Hylene W by DuPont, and IPDI (isophorone diisocyanate) by Veba Chemie, Inc.

Since aromatic diisocyanates have low solubility characteristics, they produce polyurethanes more resistant to solvent attack than those from the aliphatic type. In general, they also yield polyurethanes with higher tensile properties. Of course, the primary advantage of using an aliphatic or cycloaliphatic isocyanate is for its resistance to yellowing of the polymer by ultra violet light. In addition, these non-yellowing isocyanates have a slight advantage with regard to abrasion resistance, hydrolytic stability and outdoor weatherability. The low temperature flexibility of polymers from both types of isocyanates are about equal. Generally, the non-yellowing isocyanates are slower to react in a urethane polymerization than their aromatic counterparts. The effects of diisocyanate composition are summarized in Table 5.

Table 5. Diisocyanates Comparative Polyurethane Properties vs. Isocyanate Type

Aromatic	Aliphatic – Cycloaliphatic
Solvent Resistance	Light Stability
High Tensile Strength	Abrasion Resistance
High Reactivity	Resistance to Outdoor Weathering
	Hydrolytic Stability

The two main types of chain extenders are diols and diamines. The reactions of these two chain extenders with an isocyanate have already been described. Utilizing the raw materials described above these are combined to form the variety of polyurethane polymers necessary for textile coating.

APPLICATIONS OF URETHANE COATINGS

The tensile properties of polymers prepared from the above raw materials will now be shown. It should be noted that all of the polyurethanes that will be discussed are polyester based. In addition, the letters LS, as in LS-73 for example, denote non-yellowing urethanes. The transfer coating section will deal mainly with footwear. From this, parallels can be drawn for other areas such as apparel, upholstery and handbag applications. The second section will be a brief discussion of polymers and systems for direct coating.

In transfer coating, polymers can be roughly classified functionally as topcoat, skin coat or adhesive according to their tensile modulus properties. This can be best illustrated by the following Instron curve which shows three different polymers, each of which has been proven for the indicated application in footwear. Topcoats are normally high modulus, short elongation types of polymers which result in a rather hard, abrasion resistant coating. The skin coat must combine the performance properties of abrasion resistance and flexibility and, therefore, must have a somewhat lower modulus and higher elongation than a topcoat. Going one step further, the adhesive or tie coat is responsible for bonding the substrate to the skin. Since it is not exposed to the surface, a tough abrasion resistant polymer is not necessary. More ideally, the adhesive must exhibit high degrees of flexibility combined with maximum bond strengths and softness to impart a supple hand to the finished goods. Figure 1 dramatically points out the differences that exist as far as the tensile modulus properties are concerned for a topcoat, skin coat and an adhesive. Although these polymers have been characterized into specific categories, there is, of course, some overlapping. For example, skin coat LS 73 has also been used as a moderately soft topcoat for some applications, such as the high gloss look, where a hard topcoat cannot be used because of the heavy film deposition necessary to achieve this aesthetic effect.

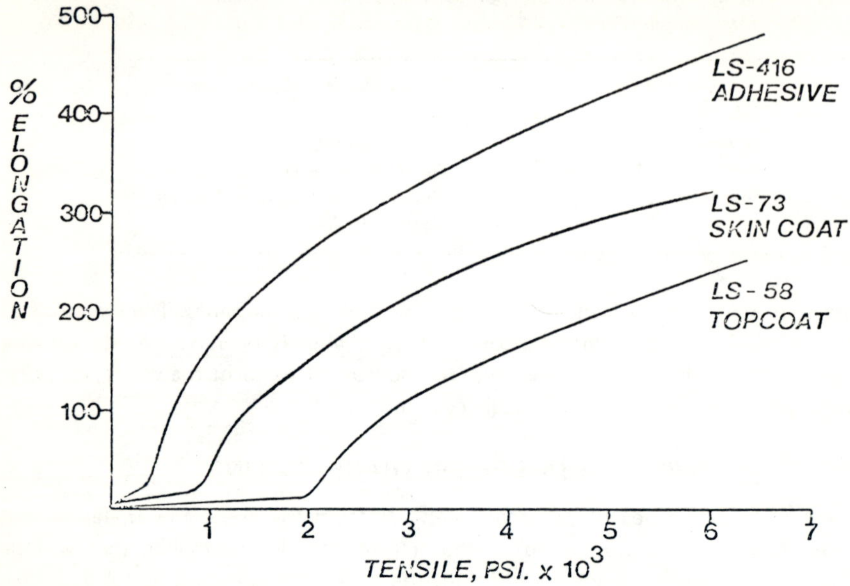

Figure 1. Tensile properties of various urethane materials.

The polymer to be used as a skin coat for a high performance application such as footwear must possess the following key properties: flexibility, abrasion resistance, solvent resistance and hydrolytic stability. Polyurethane LS 73 has been successfully used for this purpose. LS 73 is a fully reacted polyurethane lacquer containing 25% total solids in DMF (dimethyl formamide) and toluol. Initial evaluation of hydrolytic stability can be done on a free film, whereas Bally flex, Tabor abrasion and other performance tests are performed on a coated fabric composite. Results of the composite testing will be reviewed later in this paper. Free film hydrolysis of LS 73 is demonstrated in Figure 2. The film was exposed to 95% relative humidity @ 70°C for one week. It can be seen that the retention of tensile strength after this period of humid aging is approximately 96%. After two weeks of humid aging this value drops only to 93%.

Another skin that gives a somewhat softer hand is Polyurethane LS 169. This product is 25% total solids and is carried in DMF alone. This is illustrated in Figure 3. Retention of tensile strength after humid aging is 93%.

Two radically different types of tie coats have successfully been used in footwear applications. These are solution Polyurethane LS 416 and water based Polyurethane Latex 5HS. LS 416 contains 35% total solids in toluol and isopropanol. Latex 5HS is 47% total solids and is water based. Figures 4 and 5 respectively show the tensile modulus properties of these polymers. It can be readily seen that the Latex 5HS is a much softer, more extensible polymer than the LS 416. The skin and tie coat are essential components of the fabric coating. Figure 6 lists the performance values of LS 73 with both Latex 5HS and LS 146 in a typical shoe upper construction.

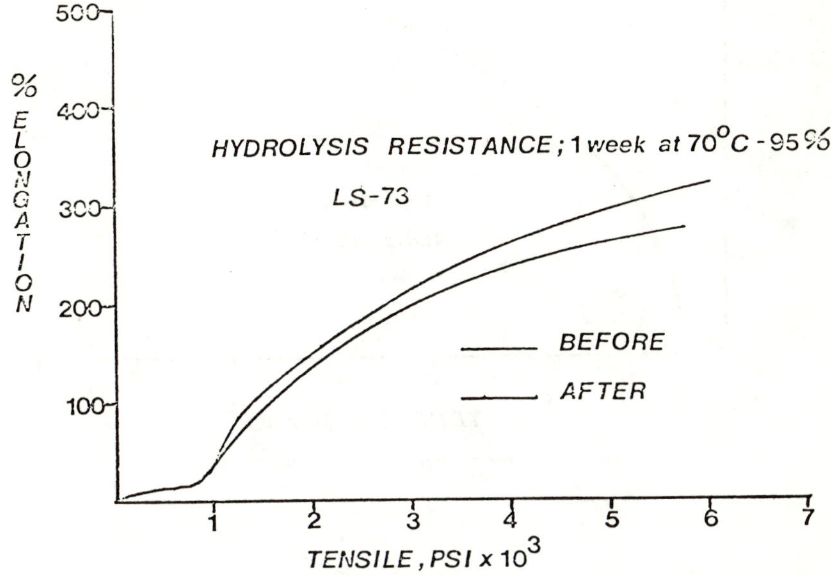

Figure 2. Hydrolysis resistance of urethane skin coat.

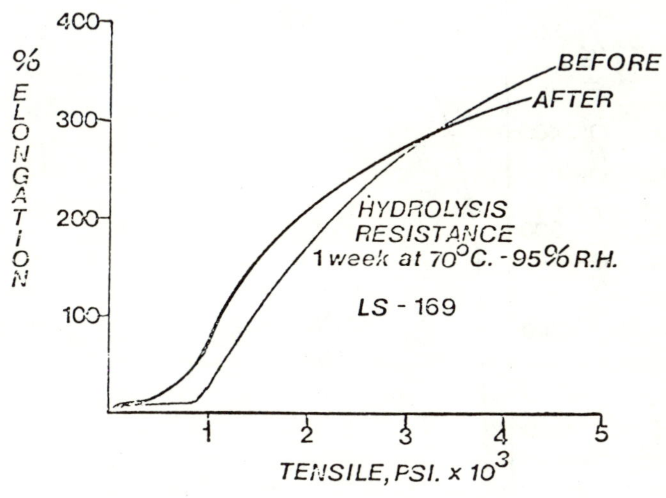

Figure 3. Hydrolysis resistance of urethane skin coat, 25% solids.

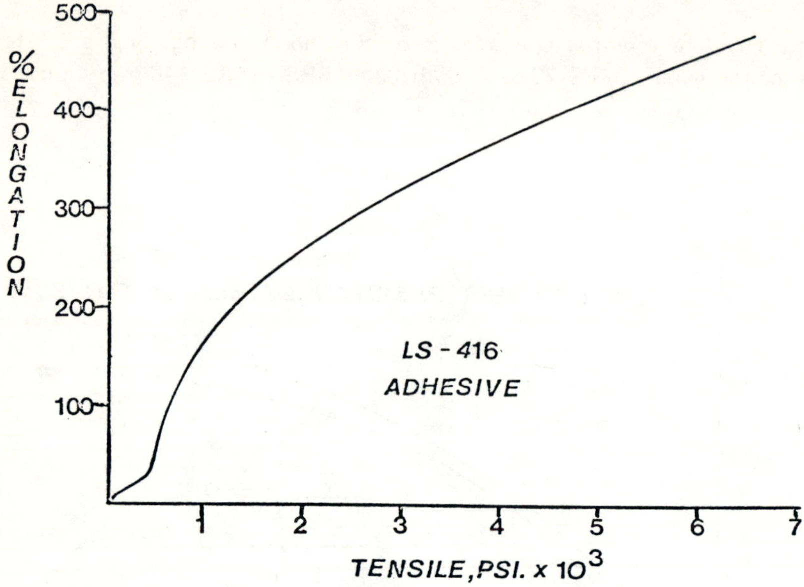

Figure 4. Tensile property of urethane adhesive.

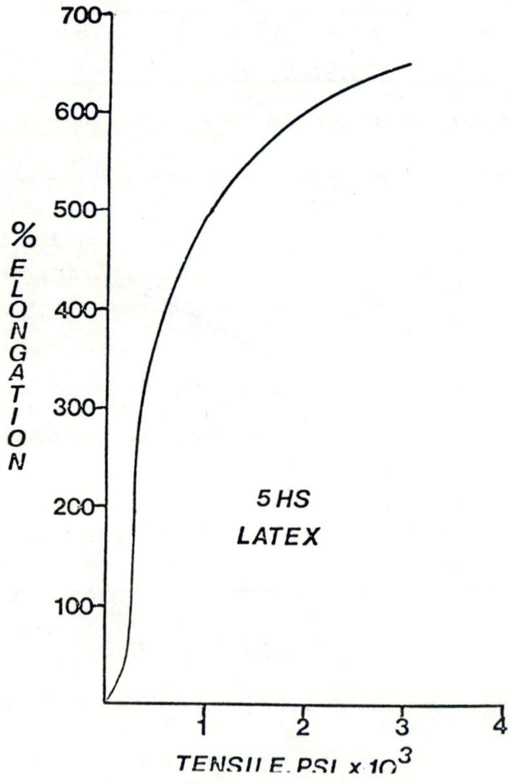

Figure 5. Tensile property of urethane latex.

CONSTRUCTION

SKINCOAT LS 73	1.3 DRY MILS
TIECOAT LS 416	1.1 DRY MILS

FABRIC - NAPPED AND SHEARED COTTON

TEST RESULTS

	LS 73/LATEX 5HS	LS 73/LS 416
Tabor Abrasion, Cycles*		
Initial	2,000	200
After humid aging**	2,300	150
*H-22 wheel, 1000 gram weight		
Bally flex, cycles		
Initial	250,000	600,000
After humid aging**	150,000	250,000
Bond strength, PLI		
Initial	20	8
After humid aging**	14	6

** Two weeks @ 70° C., 95% R.H.

Figure 6. Performance testing of footwear coated fabrics.

As can be seen from this chart, the LS 73 used as a skin coat with either solution adhesive LS 416 or Latex 5HS, surpasses the normal performance requirements for footwear. To illustrate, a Bally flex of 200,000 cycles is generally deemed more than adequate with companion requirements of Tabor abrasion (H-22 wheel, 1000 grams) of 200 cycles and bond strength of 6 to 8 pounds per linear inch. Although solution tie coat LS 416 admittedly does yield a higher Bally flex, Latex 5HS has markedly higher Tabor abrasion resistance and bond strength. Therefore, both would be likely candidates for this application. As can be readily seen, even after humid aging, both systems maintain a very adequate level of performance.

An example of a skin coat that would not be acceptable for footwear is experimental Polymer 120. Its poor hydrolysis resistance is illustrated in Figure 7.

Another type of adhesive commonly used in the footwear industry is Polyurethane 3A which is based on an aromatic isocyanate. Adhesive 3A contains 35% total solids in DMF, Toluol and MEK. This, of course, has the primary advantage of being a lower cost polymer than those based on non-yellowing isocyanates. This type of aromatic based adhesive can be used in many applications where extreme light stability is not a prime factor such as in finished goods that are made in darker colors and shades. To illustrate that the polymer fits into this area of application, tabulated in Figure 8 is an Instron graph of 3A before and after humid aging of one week at 70°C @ 95% relative humidity. Figure 8 illustrates that fact that even after humid aging, the polymer retains approximately 85% of its original tensile value.

Performance results with a coated fabric composite utilizing 3A to bond LS 73 to a napped and sheared cotton are presented in Figure 9.

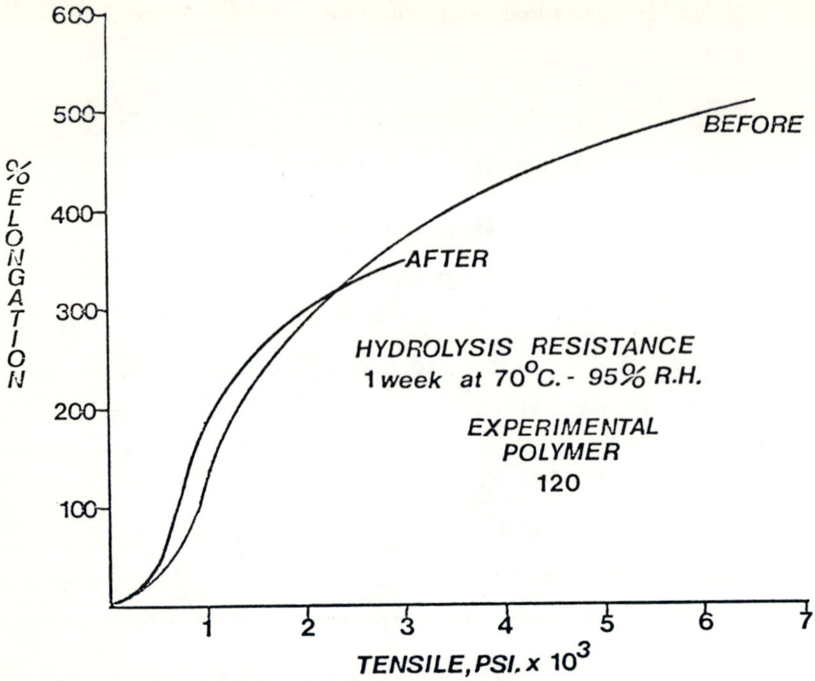

Figure 7. Hydrolysis resistance of unacceptable urethane skin coat.

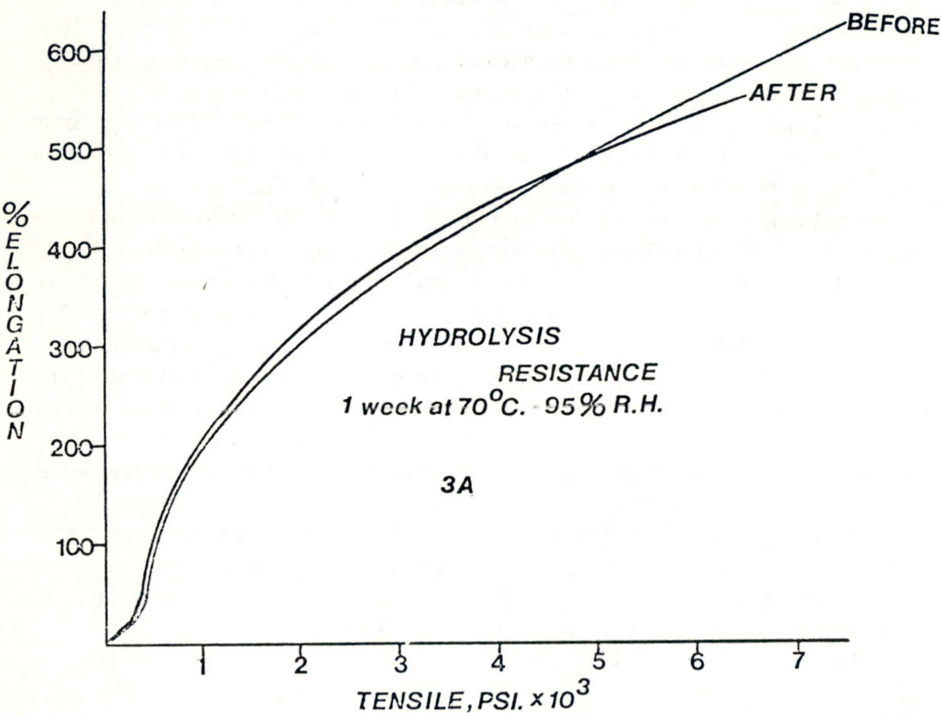

Figure 8. Hydrolysis resistance after humid ageing.

CONSTRUCTION

SKINCOAT LS 73 — 1.3 DRY MILS

TIECOAT 3A — 1.1 DRY MILS

FABRIC - NAPPED AND SHEARED COTTON

TEST RESULTS

	LS 73/3A
Tabor Abrasion, cycles*	
Initial	200
After humid aging**	150
* H-22 wheel, 1000 gram weight	
Bally flex, cycles	
Initial	400,000
After humid aging**	250,000
Bond strength, PLI	
Initial	8
After humid aging**	6

** Two weeks @ 70° C., 95% R.H.

Figure 9. Performance testing of footwear coated fabrics.

Solution polymers designed for upholstery applications must meet very similar requirements to those for footwear. Even though the types of test methods may differ, the key requirements continue to be flexibility, abrasion resistance and hydrolytic stability. Thus some conclusions can be drawn from Tabor abrasion and Bally flex. By virtue of its nature, the Latex lends to a composite a soft hand, provides high bond strengths and yet retains high degrees of flexibility and abrasion resistance.

As indicated above, somewhat different performance requirements exist for coating systems for apparel. These include very soft hand, good drape, flexibility, high slip, low tack, abrasion resistance and resistance to dry cleaning fluids.

Suitable skin coats range from Polyurethane LS 516 as a moderately hard skin, Polyurethane LS 155 as moderately soft and Polyurethane LS 190 which represents the soft end of the spectrum. These three skin coats are represented in Figure 10. LS 516 contains 25% total solids in toluol and isopropanol, LS 155, 30% total solids in DMF and LS 190 35% total solids in toluol and isopropanol.

Appropriate adhesives for apparel applications are Polyurethane LS 188, Adhesive 11-11 and Latex 5HS. These are represented in Figures 11, 12 and 13. LS 188 contains 35% total solids in toluol and isopropanol. Adhesive 11-11 is 35% total solids in toluol and DMF.

By choosing the proper adhesive and skin coat combinations, the urethane coater can achieve a vast range of softness, hand and drape. Figure 14 illustrates the performance characteristics both before and after hydrolysis and dry cleaning of a composite constructed of LS 516 and Adhesive 11-11.

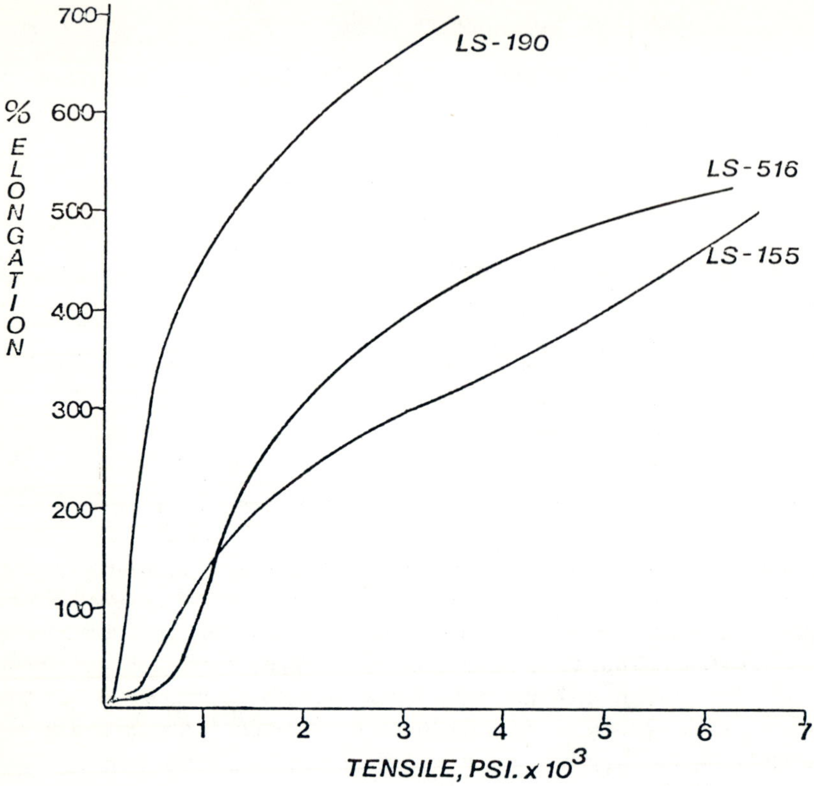

Figure 10. Tensiles of three skin coats.

A unique construction for apparel application is the use of Latex 5HS both as a tie coat and a skin coat applied by transfer with the application of a very thin topcoat. The use of the Latex in this manner greatly enhances the softness and hand of the garment, whereas the topcoat imparts an excellent wear surface combined with good slip characteristics.

With regard to the application area of handbags, cost is generally the prime consideration. Performance requirements are rather minimal. For this reason, an economical non-yellowing skin generally is selected in combination with an aromatic tie coat. Lower cost systems usually recommended for apparel are well suited for handbag applications.

Topcoats are normally used over the skin to provide an additional wear surface. They are normally applied in thicknesses of approximately 0.1 to 0.4 dry mils by gravure, knife or spray coating. Styling and design possibilities also arise through the use of topcoats when they are used for printing. Two types of topcoats, LS 58 and LS 370, are illustrated in Figures 15 and 16. Since the LS 58 is harder than the

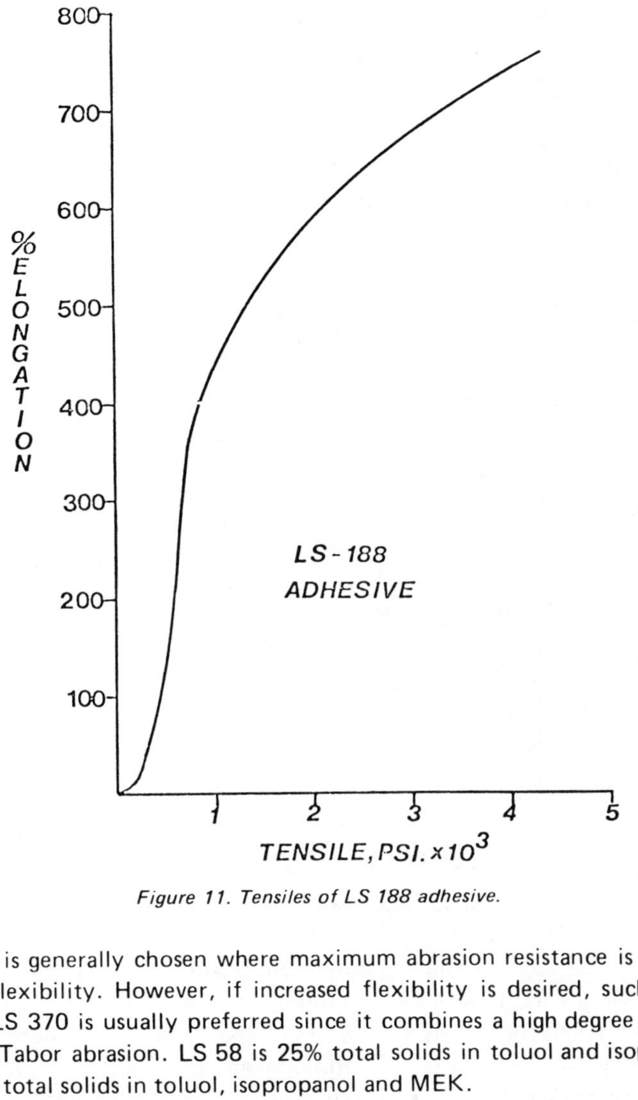

Figure 11. Tensiles of LS 188 adhesive.

LS 370, it is generally chosen where maximum abrasion resistance is desired with moderate flexibility. However, if increased flexibility is desired, such as in footwear, the LS 370 is usually preferred since it combines a high degree of flexibility with good Tabor abrasion. LS 58 is 25% total solids in toluol and isopropanol, LS 370 is 30% total solids in toluol, isopropanol and MEK.

Another major use for Polyurethane topcoats is for the topcoating of vinyl coated fabric. Polyurethane LS 370 was primarily designed to give excellent adhesion to vinyl surfaces without the need to add strong solvents or the use of prime coats. Vinyl fabrics with LS 370 topcoat have undergone extensive testing and found to possess very excellent hydrolytic stability as well as abrasion resistance and flexibility.

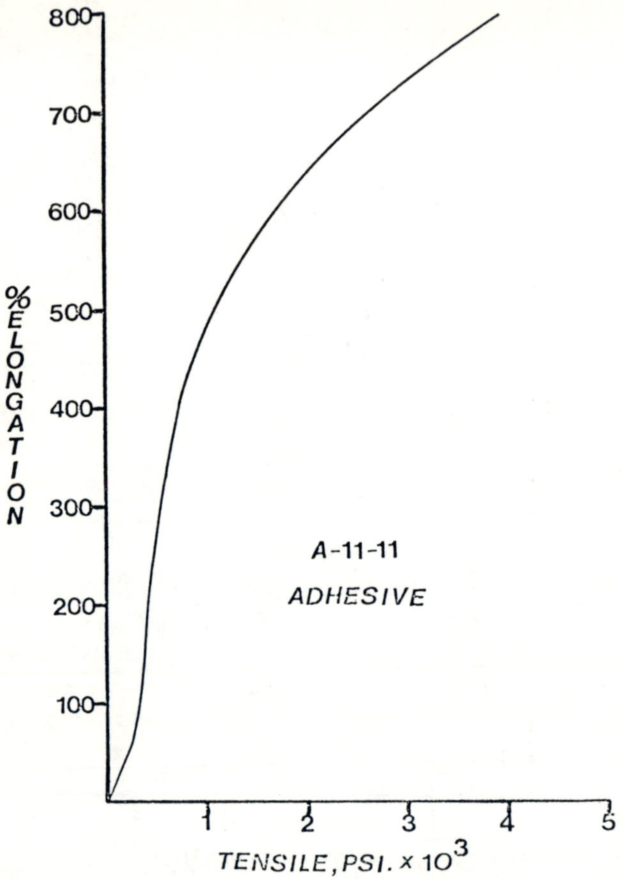

Figure 12. Tensiles of A-11-11 adhesive.

The other major use for solution polyurethanes in textiles is in direct coating. This is a field which has been growing quite rapidly and this trend is expected to continue. The major applications are the direct coating of nylon substrates for use in rain wear, ski wear and sport tents. Additional areas include inflatable goods such as airline escape chutes, life vests and military ponchos and tarpaulins. Polymers designed for this application are generally based on aromatic diisocyanates in combinations with polyesters and polyethers. Polymers such as Polyurethanes 3A, TC 95 and BC 94 have been designed for and are being used in these applications.

An important area of interest in direct coating is the use of urethane coatings that will not contribute to the flammability of the substrate. For this application, the Polyurethane FR 130 series of polymers possess the required properties. As an example, a 1.7 ounce nylon, coated on both sides with a total of 0.5 ounces of

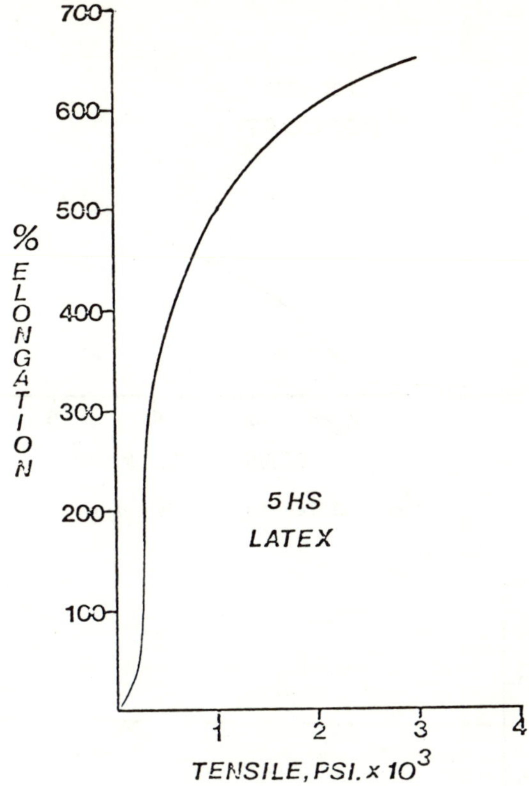

Figure 13. Tensile of 5 HS Latex.

CONSTRUCTION

SKINCOAT LS 516	1.1 DRY MILS
TIECOAT 11-11	1.0 DRY MILS
FABRIC - NAPPED AND SHEARED COTTON	

TEST RESULTS

	LS 516/11-11
BALLY FLEX, CYCLES	
Initial	130,000
After humid aging*	120,000
BOND STRENGTH, PLI	
Initial	7
After humid aging*	6
After 3 dry cleaning cycles	6

* One week @ 70° C., 95% R.H.

Figure 14. Performance testing of apparel coated fabrics.

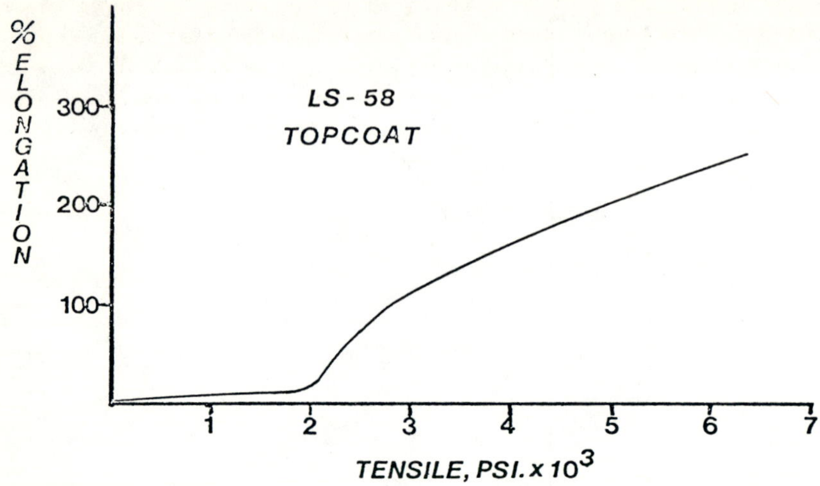

Figure 15. Tensiles of LS 58 topcoat.

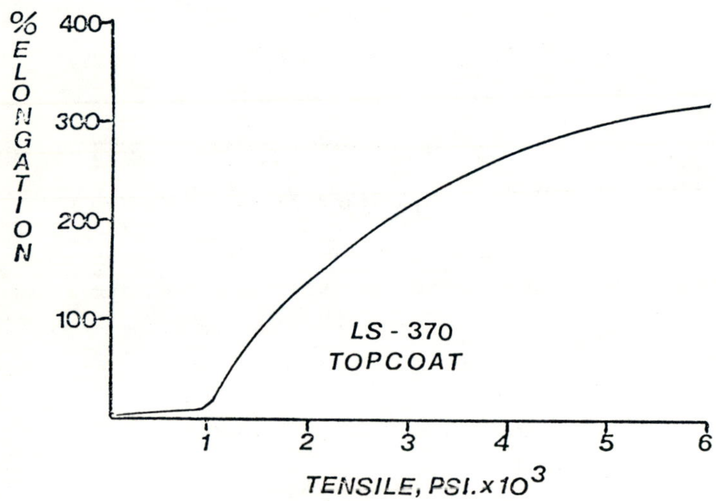

Figure 16. Tensiles of LS 370 topcoat.

Polyurethane FR 135 has passed the Federal Test Standard 191, Method 5903 vertical flame test. Approximately a 3½ to 4 inch char with no flame, was present after a 12 second burn. The use of these polymers does not affect the conventional heat sealing or air holding properties needed for their applications.

To summarize, the various urethane coated fabric markets give rise to complex combinations of performance properties. The raw materials, urethane coating systems and fabrics, pass through a variety of operations in the process of being converted to many different finished products. Art still is a major factor in all these operations. However, at each step major efforts are being made to convert this industry to a science. This paper has presented some basis for the scientific design of urethane polymers to meet the ever more demanding needs of these markets.

L. G. HOLE, ARIC, ARPS

c/o Shoe & Allied Trades Research Assoc. (SATRA)
SATRA House, Rockingham Road
Kettering, Northants, NN16 9JH U.K.

QUALITY CONTROL OF THE PRODUCT ON POLYURETHANE COATED FABRIC MANUFACTURING PLANT*

(Received January 7, 1974)

INTRODUCTION

IN A PREVIOUS paper [1] it was shown that many of the shortcomings of PUCFs in use, especially in footwear, are associated with four types of fault or weakness:

1. Fibre faults in the PU film;
2. Holes in the PU film;
3. Pronounced localised variations in the thickenss of the PU film;
4. Inadequate bond strength between the PU film and the raised pile of the substrate fabric.

The first two types of film fault lead to inadequate resistance to flex-cracking in wear, this also applies to the third type of fault with the probable additional aspect of impaired abrasion resistance. The fourth fault leads to film delamination in wear (in extreme cases during shoemaking) which results in a detached film which often becomes snagged and torn during wear thus leading to bald areas on the shoe upper or garment.

The results of the wear trial of nurses' duty shoes mentioned in the previous paper [1] provides substantive evidence for these conclusions and further support is provided by various laboratory tests.

The main aim of this article is to consider the various stages in the manufacture of PUCFs during which these faults can arise and to describe appropriate test methods which SATRA believe to be useful in assessing material quality. On this basis it is possible to suggest a possible system of "on-plant" quality control which might be applied in the manufacture of these materials.

It is convenient to consider these quality control aspects under the three general headings:

*Reprinted by permission from Shoe Materials Progress, October 1973.

Factors affecting the solid PU film;

PU film-tie coat-fabric interactions;

Fabric properties

To illustrate the various stages Figure 1 shows a typical layout diagram for a PUCF manufacturing plant and reference will be made to the various stages as indicated. Only one PU film forming stage is shown here but some plants have two, thus permitting, for example, the initial application of a transparent gloss finish layer to improve appearance and quality, prior to laying down the main pigmented coat at the second stage.

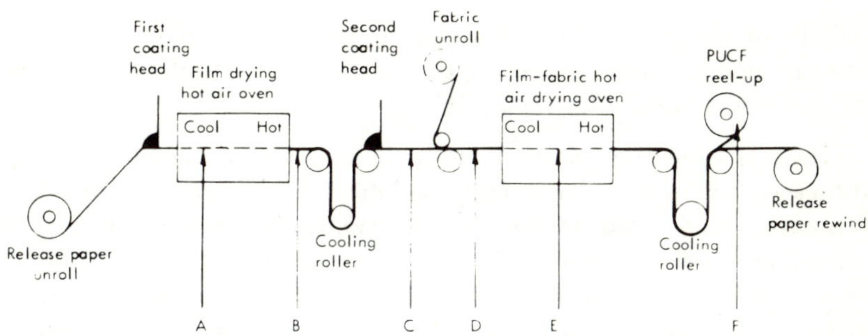

Figure 1.

FACTORS AFFECTING THE QUALITY OF THE SOLID POLYURETHANE SURFACE FILM

A variety of commercial PU coating compositions are available with a solids content usually in the region of 30%. The nature of the basic polymer varies with the different chemical suppliers but as a rule it has been of a "one-component" type which after coating and solvent removal undergoes a final curing stage which leads to a film which is crosslinked and thus will not redissolve in strong solvents e.g. DMF. This type of polymer usually has good resistance to hydrolysis – an advantage in shoe materials; however, there are with a number of polymer coatings of this type, problems of degradation by UV light and ill defined "post-curing" changes both leading to film embrittlement and cracking in service. Various solvent systems are also used by different composition suppliers although these are usually based upon mixtures of ketones, esters and aromatic hydrocarbon with the possible addition of a proportion of DMF – solvent composition is dictated by the solubility and coating characteristics of the polymer.

The three important stages for the application of the coating composition is indicated by points A, B and C in Figure 1.

Points A and B

A thin film of the approximately 30% solids coating solution is applied to the release paper by the first coating head and rapidly passes into the first oven. Incorrect oven temperature and air circulation at point A, a stage when the coated solution is losing solvent rapidly, can lead to skin formation and the creation of bubbles in the film due to solvent vaporisation under the skin. In extreme cases the bubbles at point B in the almost dry solid film may be obvious thus leading to rejection, but usually the film may appear sound to the eye but in fact have many pinholes due to bubble formation in the oven. Thus film samples may be taken at point B and assessed:

(i) visually by transmitted light using a low power microscope;
(ii) strength and extension properties of ring specimens. (Both of these techniques are described in reference [1])

On-plant visual assessment might be made by a trained operative in a minute or two and the ring test in a few minutes if a suitable testing machine is close-by. The former inspection could probably be done continuously on some coating plant by a closed circuit television monitor with microscope optics (cost about £400), thus not damaging the film by sample removal. The ring test might be carried out periodically e.g. the beginning and end of a fabric reel, since sample removal is clearly essential.

Point C

When the surface film reaches the second coating head it is approximately 0.05 mm thick and is largely free of solvent and having been cooled over rollers, it is in reasonably strong condition. The second coating head applies the polyurethane adhesive solution or tie-coat; the means of bonding the film to the fabric substrate. These tie coat compositions are often of the "two-component" polyurethane type and after drying tend not to be crosslinked to the extent that the surface film is. As a rule they have relatively low softening and melting points and are susceptible to dissolution in ketones and strong solvents, a potential problem for some stages in shoe manufacture. They are also more readily hydrolysed than one-component polymers by moist heat especially in the presence of catalysts, e.g., weak bases. Thus the tie-coat polymer may lead to quality and performance problems during the manufacture and wear of shoes; however, it might be noted that alternative polyurethane polymer tie-coat systems, i.e., based on aliphatic isocyanates rather than the aromatic types, are now under development and in use to minimise or eliminate some of these shortcomings.

The relevance of the tie-coat application at point C to quality control concerns the following effects:

(i) solvent from the tie-coat being absorbed by and softening the surface film; this can lead to pronounced crazing effects in the film when tie-coat solvent systems have not been properly formulated;

(ii) partial dissolution of the surface film by the tie-coat solvent leading to a weakened film of varying thickness (see Figure 2 [1]). A probable concomitant effect is the intermixing of the film and tie-coat polymers with results very likely adverse to the physical/chemical requirement of both.

Quality monitoring may be done by taking samples at point C and applying visual assessments for dissolution and crazing effects.

A ring tensile/extension test might also be done to provide supporting data after superficial drying of the tie-coat, also to assess the degree of softening caused by solvent absorption.

Modifications to tie-coat formulation may be indicated by quality checks at this point. A major factor at this stage, however, is that of solvent softening of the surface film and the relevance of this to the fabric combining process.

Reticulation (network patterns) effects have also been observed by transmitted light on the fabric side of the surface films of some commercial materials – the appearance being that of a network of thread-like asperities thus suggesting as a possible cause the method of application of the tie-coat solution to the film. On the other hand this reticulation on the fabric side of the film might also be due to limited solvent crazing such as mentioned above. Whether or not reticulation effects of this type on one side of the film have a significant effect on PUCF durability in service is not clear.

INTERACTIONS OF PU FILM-TIE COAT-FABRIC COMBINATIONS

Over-softening of the surface film by tie-coat solvent can lead to two important adverse effects during passage through the fabric combining nip.

(i) permanent penetration of the fabric pile fibres deep into and through the surface film. The former is undesirable because of the weakening effect on the film – the latter is further unacceptable because of the impaired surface appearance and feel of the film. Complete fibre penetration was more of a problem with direct coated PUCF than with the transfer process.

(ii) temporary penetration of fabric pile fibre loops into the softened film – after leaving the combining nip many of these fibres detach from the film but leave a permanent imprint of their temporary impression into the film.

This type of fault is perhaps the most important since controlled wear trial experiments and laboratory tests have shown that pronounced "fibre-faults" in a PUCF will lead to premature flex-cracking in wear. Moreover fibre-faults are not usually readily detected and are certainly not visible when the material is viewed on the "grain" surface by reflected light. These faults were first demonstrated in Reference 1 by examination of the surface film by transmitted light – it being necessary to first remove the PU film from the fabric for which operation a leather splitting machine was found to be very effective. Samples for the assessment might be taken at points D and F. Subsequent experience has amply supported the conclusion that fibre-faults create a serious weakness in material which in their absence might otherwise be a satisfactory product.

Clearly solvent softening of the polymer is not the only single factor in the formation of fibre-faults – others deserving of close control or observation are:

combining pressures in the nip
time between tie-coat knife and the nip, i.e., machine speed
material temperature at the combining nip
nature and quality of the pile on the fabric
fabric feed direction and angle of incidence at the combining nip.

Unfortunately it does not seem practicable to monitor for fibre-faults by a non-destructive method. Samples must be taken from the combined material and the film separated from the fabric by splitting – the film then being examined by transmitted light under a low power microscope.

Two further quality control tests on the film samples at points D and F relevant to fibre-faults are:

Ring tensile/extension tests (time scale – a few minutes)
Cyclic biaxial extention (dome) test (time scale – 5 days).

"Bubble Crazing" effects is a fault which apparently can arise at point E in some coating processes. The basic cause is similar to that which arises at point A – solvent in the tie-coat vaporises too quickly if the oven temperature is too high in the initial stages – likewise the PU surface film is softened by both the temperature and the solvent. The result is that a pattern of relatively large bubbles rises all over the material in a craze pattern – the junctions between the bubbles being the lines of remaining adhesion to the pile fabric. When the material emerges from the oven the lower temperature causes the bubbles to collapse and some measure of adhesion to establish itself. The craze pattern of junction lines remains, however, and apart from general material shortcomings such material, although having a characteristic appearance, is not attractive. It is not always easy to balance the various requirements of the process to eliminate this unacceptable fault. In difficult cases the polymer and solvent systems may need to be changed to eliminate the problem.

FABRIC QUALITY

It is well known in the coating industry that the uniformity and general quality of raised pile fabrics is the major single factor which influences the general quality of a PUCF. One coater has stated that to maintain acceptable quality he has been obliged to reject 50% of his fabric deliveries – whatever the exact figure, there is no doubt that fabric variability can be the cause of serious production difficulties and make difficult the manufacture of PUCFs of consistent quality.

In SATRA's experiments it had been expected that some differences in peel-bond strength between film and pile might be due to with-and-against pile direction but unexpectedly the greatest difference observed has been across the pile, i.e., across the roll, a result which does not have a very obvious explanation – assuming that the combining nip is properly adjusted.

Quality control methods with pile fabrics would seem to relate to:

(i) *Uniformity of pile length.* In this connection some fabric manufacturers are known to shear the pile to uniform length – others are thought not to do this. Sheared fabrics of good quality are generally more expensive – thus as with most things, quality demands a higher price and clearly some coaters are not prepared to pay it.

(ii) *Damage to the pile.* It can be demonstrated in PUCFs that delamination of the film in some materials is associated with a high proportion of fibre pull-out from the woven fabric. This may be due to the raising process having caused damage to the fibre "root" but it may also be due to low quality cotton of shorter staple length. No matter what the causes a simple peel test provides the evidence for the nature of the failure, i.e., a large proportion of fibres adhering to the PU film in materials with low peeling loads.

(iii) *Direction of lay of the pile.* A peel test in the two principle machine directions (with and against) will provide a quality check – however, the result might indicate a rewind of the reel(s) as being desirable. Clearly some control may be possible by varying the angle of incidence between the fabric and the PU film-paper laminate at the entry to the combining nip.

Thus a valuable quality control test for raised fabrics is the peel test as described in Reference 1. This presupposes consistent PU film and tie-coat properties as well as combining conditions; however the test might be modified by using a consistent substrate e.g. a metal bar, with an adhesive system and laminating technique providing constant bonding conditions.

SUMMARY OF QUALITY CONTROL TESTS

1. Visual examination of the PU film by transmitted light with the aid of a low power microscope. An inexpensive student's grade stereoscopic microscope with X20 magnification is adequate for the purpose. Fibre-faults and pinholes will be absent in a good quality film of 0.05 mm thickness.

2. Load-Elongation tests on the PU film using ring samples. Extension at break should be over 200% and the desirable range 300–400%. (The amount of solvent in the film at test will influence extension and correlations should be made with mature solvent-free film having the extension values indicated.)
3. Peel Test: Combined Material. The peeling load of the PUCF should not be less than 1.5 kgf/cm width in the principle fabric directions. Green strength values may well be lower and appropriate correlations should be made with mature material as in (2).
4. Peel Test: Raised Fabric. The use of consistent substrate, e.g., a metal plate or a sheet of resin rubber, has been suggested above and peel loads of not less than 1.5 kgf/cm width should probably be aimed for in a suitable raised fabric. It must be added, however, that we have had no experience of this modification to the peel test for PUCF but we expect to publish some data later for a range of raised fabrics.

Finally, it is recognised that a number of other very important quality control aspects have not been considered, for example the nature, viscosity, wetting and flow properties of the PU coating compounds and the influence of pigments and other additives. As coaters well know a number of other aspects of control are essential – all that this paper has attempted to do is to draw attention to the known shortcomings in some of the more significant mechanical properties of PUCFs and to suggest how these properties might be monitored and controlled on a PUCF coating plant by applying simple tests.

REFERENCES

1. D. Popplewell and L. G. Hole, Shoe Materials Progress, August 1972, pp. C1–C15.

H. W. Leigh
Consultant

Urethanes Application and Technology
1 NAB CLOSE
Bollington, England

DEVELOPMENT OF URETHANE FOAMS IN TEXTILE LAMINATION

(Received January 7, 1974)

INTRODUCTION

To introduce this paper on development in polyether and polyester foams for textile bonding, I should like to give a short history of the subject of urethanes generally to put the developments into perspective. In 1937 Otto Bayer and colleagues began to develop urethane technology in the sense of the basic reaction between isocyanates and amines, hydroxyl materials, etc.

During the war years both Germany and America were working on rigid foams (based on urethanes) and also lacquers, paints and adhesives. In the early 50's flexible polyesters based foams were developed in Germany by Bayer where the emphasis was on producing, in a continuous fashion, flexible polyester foams on a large scale. The latter end of the 1950's saw the introduction of the cheaper polyether urethane foams produced via prepolymers and around about the same time foam slit into thin sheets was finding application in textile bonding to give laminates with textiles, conferring excellent insulation properties and stabilizing the textile used. The foam produced at this time from polyesters tended to be rather stiff but nevertheless they found a ready market and a marked development occurred particularly in America, both in the polyester used in the production of foams to produce softer and better draped materials and also in the process of lamination.

The late 50's and the early 60's saw the introduction of the one shot polyethers and also that of the flame lamination process patented by Curtis Wright of America and now owned by Reeves Brothers. The period between 1960 and '62 to '63 saw a rapid increase of lamination in both America and the Continent and also the UK but the understanding of the lamination processes the types of textiles which could be usefully and successfully bonded, the finishes, the general quality control of the lamination process was suspect. Consequently a disillusionment period set in, in the early 60's to about 1964 when people lost faith in the product and the product itself was rather suspect in that claims made for washability, dry cleanability, were not always justified.

About this time the raw material suppliers, the textile laminators and the foamers were working towards improving the qualities of foams and of lamination and seeking a better understanding of the processes. The round block production of urethane foam was developed which allows a peeling operation of a more efficient nature and greater lengths of peeled foam sheets could be produced. The lamination processes were improved throughout the middle to late 60's and a progressive improvement in the technology, the understanding of urethane foam production to meet specific and very tight requirements was obtained. A new stabilizer was introduced which improved the control of cell structure in foam, which is a fundamental feature of drape and handle characteristics and to some extent also the lamination characteristics of the material. A better understanding of the mechanism of foam bonding, the finishes of textiles which gave improved or reduced bond strength of laminates was being gained and this period of the latter 60's and now into the early 70's is one of understanding lamination. Joint programmes are being undertaken between laminators, foamers, raw material suppliers and customer, in conjunction with academic institutions and textile and associated research establishments such as the Shirley Institute, the Dyers and Cleaners Research Organization and University of Manchester Institute of Science and Technology.

From this outline then, it can be seen that the developments of urethane foams for textile use have taken place over a relatively short period of time. It is in this atmosphere of rapid development, rapid increase in consumption, marked improvements in quality of products perhaps that I wish to place my paper.

CHEMISTRY

The three basic units involved in the manufacture of urethane foam are a polyol which can be polyester or polyether, an isocyanate and water. The principal reactions taking place are:

1. That between the polyol and the isocyanate to give the urethane linkage, which because it can build up in a continuous fashion becomes named or mis-named a polyurethane unit or molecule and
2. The reaction between isocyanate and water to give CO_2, the blowing agent creating the foam.

Ester (Produced from diol and diacid condensation).

$$\mathrm{HO\cdots OH + HOOC\cdots COOH \quad HO\cdots O{-}\overset{\overset{\displaystyle O}{\|}}{C}\cdots COOH + H_2O}$$

Ether (Produced from polyhydroxy compound plus alkylene oxide).

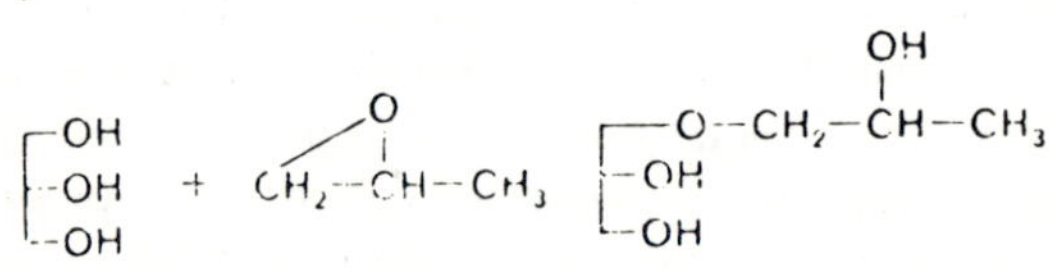

Polyhydroxy base (eg Glycerol) — Propylene oxide

Reactions

(1) Isocyanate + polyol ——→ urethane
eg

$$-NCO + HO- \longrightarrow -\underset{}{N}(H)-C(=O)-O-$$

isocyanate polyol hydroxyl group URETHANE

(2) isocyanate + water ureas + CO_2 (gas)

$$-NCO + H_2O \longrightarrow -NH_2 + CO_2 \uparrow$$

$$-NH_2 + NCO \longrightarrow -N(H)-C(=O)-N(H)-$$

SUBSTITUTED UREA

In practice mixed structures occur.

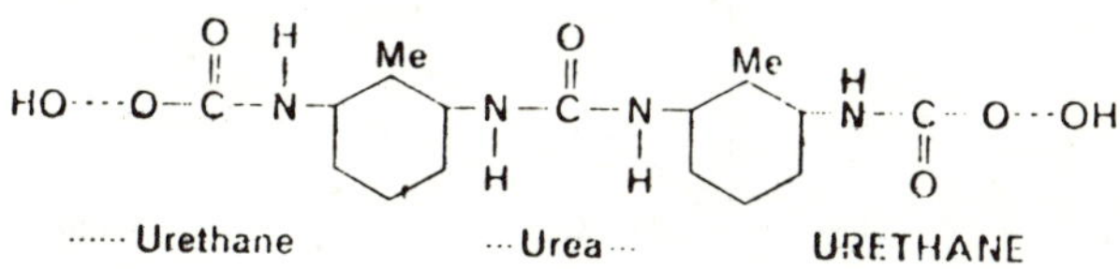

Cross-linking

Type (i) Increase T.D.I. level—foam gets harder.

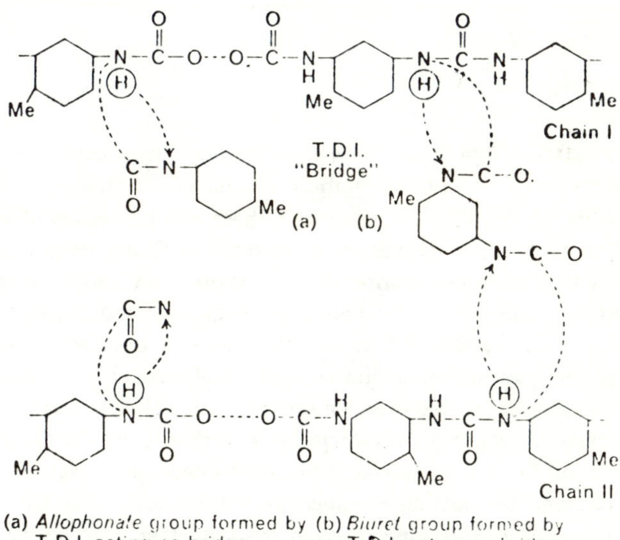

(a) *Allophonate* group formed by T.D.I. acting as bridge between Chain I and Chain II by reacting with active hydrogen (H) of urethane group in each chain.

(b) *Biuret* group formed by T.D.I. acting as bridge between Chain I and II by reacting with active Hydrogen (H) of substituted urea group in each chain.

Type (ii)

Branching available in the polyol used.

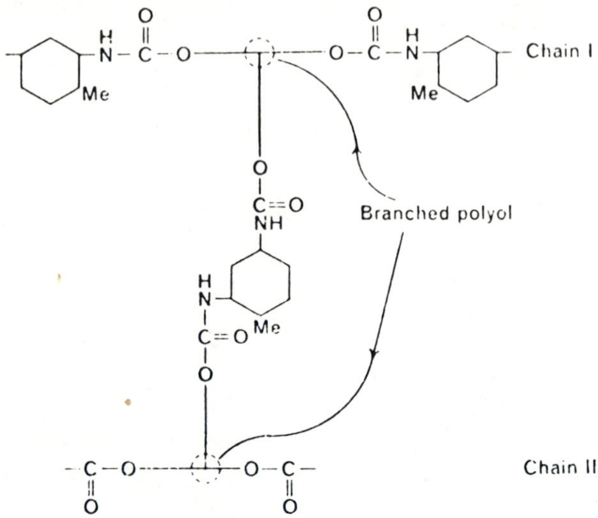

Chain I and Chain II are cross-linked through a "bridge" formed when T.D.I. reacts with the reactive hydroxyl group of the branched polyol unit within each chain.

Type (iii)

Low molecular weight polyfunctional cross linkers can be used.

Since the reaction rate is relatively slow, to push the reaction along catalysts are required. In the case of polyesters these are tertiary amines of various types and varying activity. In the case of polyethers these can be blends of amines and also heavy metal salts or soaps such as stannous octoate. Since the bulk of lamination is based on polyester for the purpose of illustration, I will talk only about polyesters at this point. As you can see by taking an isocyanate group and then a polyester group a further isocyanate and ester which gives a urethane linkage, water and isocyanate which gives a urea or substituted urea linkage and so on along the chain, nevertheless the product the way I have shown it is linear and this as you would reasonably expect has got good elongation characteristics and depends upon stiffness primarily on limited crosslinking between linear molecules of the type I have depicted. However, by putting a branch or a third reactive group in the polyester chain we can create crosslinking through the polyester itself so by placing a small side chain on the polyester I can get a further isocyanate in position and this will stiffen up the molecule.

These two types of polyester, the so-called linear or difunctional polyester and the branch chain polyester, three functional, are the two basic types of polyester used in flexible foam production today. The linear type is used primarily in textile grade foams whilst the branched type is used in the so-called industrial grade foams where a stiffer characteristic is required in the product and high elongation, high tear strength, and high tensile strength are not necessarily a prerequisite of the product. We now have the three major components, we have catalyzed them and are making a foam, the viscosity balance the viscosity differential between the three major components is very wide, the polyesters conventionally used in textile foam manufacture range in viscosity between twelve and twenty thousand centipoise whilst water and isocyanate are very low, round about five centipoise for TDI for example and a similar viscosity for the water, catalyst, surfactant blend.

These materials need to be rapidly mixed to be allowed to react and then stabilized using special surfactants. It is possible, using blends of branched and linear ester resins, to produce materials of a wide variety of stiffness characteristics. For example, here we have two samples about the same thickness of foams which are in the first instance good quality textile laminating quality foams for mini foam bonding. As you can see over the edge of the table, this foam draped quite markedly. On the other hand a more heavily crosslinked polyester resin coupled with the same type of isocyanate would give a foam which under the same test stands out much more proudly having much more stiffness within the unit. This stiffness can be attained in several ways, in the molecular sense as outlined by the branching characteristics but also this can be attained by the cell structure control so that the larger the cells produced in the foam the more stiff the foam becomes: the finer the cells the softer in handle and the better the drape of the foam.

We have now drawn the picture of the area in which the textile foam bonding process developed, the sort of time scale which has been involved and the rough outline chemistry of what one needs to do to produce a foam. It has been said, producing foam is easy and this in fact is the case. To take all the components I have mentioned and mix them in the so-called hand mix technique for a period of perhaps six to ten seconds creates quite a reasonable emulsion. On pouring this into a cardboard box, the foam rises and some 90 seconds later you have a piece of urethane foam. It looks good, it feels good, until you cut it open and then you find that because your mixer wasn't quite at the bottom of the vessel or you didn't quite scrape the sides during stirring your mixing is uneven and the product has got coarse cell patches and streaks.

We can now move onto the process involved in the manufacture of foam and the effects of the various parameters on the quality of product and evenness of product and the demands of customers which influence the product level on plant and the quality control of the foam involved.

PROCESSING

If we show the three major components of the foam reaction in the form of blocks then this is a reasonable simulation of the proportions going into a normal polyester foam mix that is to produce a foam of, say, 26–28 kg/m^3 density. The viscosity of the components are marked along the side (I discussed them earlier), we have the problem now of mixing these proportions of components, all of which are pumped into the mixing chamber separately, and we have to get even mixing, and good dispersion. As you can see from these minor components like water, catalysts and surfactants they are small quantities compared to the bulk polyester resin or the isocyanates. We need to control the reaction so we need to control the catalyst level accurately and the temperature of the major components and we like to control them within, plus/minus one degree centigrade.

We are now pumping controlled temperature chemicals in the form of the major components through a mixing chamber carrying out mixing and we want to discharge as evenly and smoothly as possible onto a moving paper belt to allow the foam to rise evenly, without being disturbed unduly during the rise, to form into either a round or square shape. In the high pressure machine, which is the one we ourselves use, the bulk component, the polyol (the polyester resin that is) is pumped at fairly low pressure as the main stream component and pumped directly into the head through an open valve as shown. All other components with the exception possibly of dye paste are injected at high pressure into the head through injectors operating anywhere between 50 and 180 atmospheres pressure on the high pressure side of the valve, of course, into (a simple peg stirrer) and mixing head where the pressure is conventionally only of the order of one or two atmospheres. The material now needs to flow through this mixing chamber and be thoroughly mixed. The mixing, because of the spray action of these components can be good, very good or even indifferent it depends upon the placing of the injectors the operating efficiency of the injectors, the spray characteristics that one gets, the choice of injector so that the efficient mixing is attained as much as by the physical dispersion of the material through the injector spray effect as it is by the swirling motion and the mixing motion of the spinner itself. In fact, in a high pressure machine it is quite possible to mix components efficiently perhaps not at such widely differing viscosities but nevertheless efficiently without use of a mixer at all.

Having controlled the temperature of the major components and mixed them thoroughly in this mixer, we have to obtain an even smooth flow throughout the mixer since we do not wish to have dead spots because these cause build up of raw materials which can give pre-nucleation of the gases within the mix which is coming down through the mixer and this can cause premature bubble formation in the mix at a too early stage and this in turn can give rise to large holes or bubbles within the foam. Any dead spot area in the mix or in the nozzle, any build up of components during the course of a normal production run can lead to defective foam in the form of holes, striations, so-called water marks, etc.

CELL SIZE CONTROL

Control of cell size in any foam formation is primarily one of mixing, nucleation and stabilization. The more atomization of your various components you get, so that the droplets are finer, the better the nucleation. In the case of a high pressure head the lower the head pressure creating a bigger differential between the injected components and the major components the finer the cell formation. In the case of low pressure machines a conventional improvement of cell structure in terms of improving the fineness of structure is reasonably readily attained by increase of mixer speed and also increase of air rejection, the air acting as the nucleation centre for the gas formation reaction in the foam.

FOAM DEVELOPMENT AND SHAPING

The foam mixture is now poured onto a moving paper belt and the demand at this stage is for even distribution of the foam mix with a steady state condition applying as soon as possible on the paper band. The foam at this stage being delicate enough for it to need looking after. In the case of round block production there are specific considerations because the reaction rate and the moulding operation into the round form have rather critical limits so that catalyst variations and conveyor speed are somewhat restricted.

The foam now rises up to take its final shape which can be round or square block. The effect of the nozzle in the lay down is very significant. We ourselves have taken a great deal of pains in this particular area to avoid the dead spots I spoke of earlier, premature nucleation giving holes in foam is the bane of the textile foam manufacturer's life.

IMPURITIES

Any physical impurity in the foam chemicals such as crystal formation in the isocyanate, oil, greases, silicone oils, general impurities of a nature which will either cause foam breakdown because they are anti-foam materials like oils and greases or physical bodies which act as large nucleation sites and consequently create massive bubble formation which can create holes in the foam are to be avoided. Having finally got the mixed chemicals onto the paper we don't know whether the foam is good, bad or indifferent. It might look good, the reaction line is fine and the angle of rise is fine; it fills the mould at the right stage, the "bubble off" of gas is where we expect it, but we still don't know whether we have got a very good dispersion of all components, whether we have got any holes in the block for example. The first stage then is the "chop" at the end of the conveyor line. A sample is taken at this stage peeled rapidly and then a visual examination is made, for holes, streaks and striations. If this is acceptable then the run continues.

QUALITY CONTROL

Checks are made progressively throughout the run and if something unusual is noticed adjustments are made if possible, if not the run is stopped. With mini foam bonding, i.e., the lamination of fabrics using very thin foam films, small continuous streams of holes or large single holes, which could be quite acceptable in substantially thicker gauges of peeled foam or slit foam, are critical. The flaming process can burn through in these areas and the whole foam band can tear or if adhesive lamination is being carried out the adhesive can penetrate the holes and give a stiff, poor handle product.

Samples of the production run are checked in the laboratory against specification which can be British Standard, foamers own, or special specifications agreed with the customer. Having passed the tests for colour, cell count, hardness density, tensile strength, elongation, porosity, resilience, etc., the foam is peeled or slit to customers requirements.

PEELING

After being cleared of all laboratory checks the foam is now sent for peeling, the next section therefore is obviously the peeling operation, the problems of peeling the range of thickness and widths demanded by the customers and the quality that the customer requires to meet his own critical limits for bond strength, drape, dry cleanability, etc. In the early days of the peeling operation widths up to 54 to 56 inches were standard and slowly the demand grew for widths progressively wider up to the now fairly common widths of 66 and even 72 inches. These developments have been associated primarily with the increased widths available from knitting machines and as these have gone up so have the demands on the peeling operation. Peelers are now available which will operate up to 125 inches and even more or though at this point in time the knitted fabrics or woven fabrics for that matter are not available which require such widths of peeled foam. The demands on the peeling section are such that the quality of foam must allow for an evenness of peel and evenness of feel and handle so that the customer in lamination whether by the adhesive method or the flame laminating process can obtain a good quality reliable and reproducible product.

In the early days of foam bonding a thickness of the order of 2 mm was about as low as one would normally be requested for, over the last few years we have seen the growth of the so-called mini foam process in which approximately 1.5 mm of foam and even less is burnt up almost entirely so that the foam layer becomes only the adhesive between the two materials and a wisp of foam is left in between the two fabrics. Since in adhesive lamination there is no reduction in thickness foams of even lower gauge have been requested over the years and now the competent foamer and the textile foam supplier must be in a position to supply consistent quality, consistent thickness down to 0.5 mm of foam which is rapidly approaching the limit of cell size for the average textile foam and in fact it is below the cell size of some of the coarser grades of foam. The prime requirements of the laminator are that the tolerance of the foam supply should be within tight limits. For example the recent specification BS 4021 which was drafted and pushed through in the main by a combination of foamer, peeler and laminator as well as end user.

Because of the requirements of the laminator that the foam quality in a particular roll length should not vary by more than —0 + 1 mm for thicknesses up to 2 mm, this means that the bulk of all lamination these days requires extremely tight control, particularly throughout the course of a roll length and variations must be minimal. This has been an area of marked development so far as we are concerned since the location of the peeling machine, the angle of the land on the knife, the type of knife used, the effect of joins in knives, speed of running, alignment of the nip pull, whether one has grinders on or off, etc., have all formed part of a long term development programme to produce the required products which now operates quite successfully at 0.5 to 0.6 mm and widths up to 72 inches.

DEVELOPMENTS

It will, I hope be apparent from what has been reported so far that the major developments in the areas of urethane foams used for textile bonding over the last five years have come primarily from the foamer in improving his quality of product, understanding better the mechanism of production to give a more evenly structured product, more regular control of physical properties, better quality of peeling, different colorations, wider blocks, more continuous lengths, round block developments, etc., rather than from the raw material side. The actual raw materials used now with the exception of the silicone surfactant foam stabilizer which was mentioned earlier have been in use for at least 8 to 10 years in the case of polyester textile foam and one naturally looks forward to new chemicals in the hope that these will help to solve some of the problems which still face the foamer and the laminator. However, whilst the foam producer knows his equipment and operating conditions better than the raw material suppliers, he depends on his suppliers for that special degree of knowledge which only they have.

As outlined much of the development work in processing foam for textile lamination has been directed towards a better understanding of the parameters involved, raw material suppliers have not stood idly by. It is difficult to know which way to modify a polymer until the basic factors affecting its use are understood – at least in part.

DEVELOPMENTS IN RAW MATERIALS

All flexible polyester and polyether foams can be laminated by using adhesives but only polyesters and special polyether based urethane foams will laminate by the flame process. The melting point range and adhesive properties of the partially decomposed ester foam is suited to the flame process but conventional polyether foams degrade under heating over a narrower band and the molten product possesses poorer adhesive properties. Also the ester foams are intrinsically stronger having higher tensile strength and in the case of textile grade foams higher elongation than the "equivalent" ether based foams.

It has been known for many years that adding fire retardant additives to polyether foams conferred flame welding properties. These additives are non-reactive but appear to "spread" the melting point range of ether foams and also modify the adhesive properties of the molten foam. From the early 60's products have been offered based on polyethers and a market has existed but since any wetting of the foam bond particularly by dry clean solvents weakens the polyether textile foam much more so than that of an ester foam, even with some price disadvantage polyester textile foams are the principal type used. Reactive additives are now available which when introduced into a normal polyether foam formulation will give flame weldable foams having reasonable dry peel bond strengths.

I mentioned earlier the expansion of colouring in foams to give a very wide selection to the laminator and his customer. We also have the need for white foam and since large foam blocks tend to discolour in the centre as block sizes have increased so the demand for a better control of "scorch" has grown. For a standard formulation producing easter foam at say 25 kg/m^3 in blocks some 30 inches high scorch in white (unpigmented) foams present no major problem. However, as the block size increases so the core will show, the same formulation, an increase or development in discolouration.

Making, as we do, large round blocks over 40 inches diameter created a problem which, by a combination of formulation modification and co-operation with raw material suppliers, has been overcome. The two sections of block show this change, the discoloured block being improved progressively to give the product at present made which shows much reduced scorch.

This illustration of joint effort to overcome specific problems is not isolated and serves to show that whilst foamers are not being supplied with new "solve it all" products regularly progressive improvements to raw materials are being obtained.

PROCESS DEVELOPMENTS AND UNDERSTANDING

An area of major development is that associated with an initiated by customer demands where different characteristics of stretch, of bond strength, handle, colouration, post dyeing and printing, cell size control etc., all stimulate the producer to modifications and progressive trial and error testing with the customer. This is the area of development at the present time. A major effort has been made over the last two years to improve our understanding of the lamination process. Work carried out by DCRO in conjunction with the laminators, the end users, the fabric suppliers, pointed to a fairly complicated situation. In conjunction with Shirley Institute the foamers, laminators and end users have co-operated to produce some rather startling information. For example it was considered that the prime mechanism of bonding was that of mechanical bond. It has been proven recently by members within this joint group that a bond of foam by the flame lamination process to a cast sheet of nylon can, properly done, give a similar bond to that of a woven or knitted material of the same basic composition. When you know why things happen then you are in a position to control the situation. Knowledge of fundamental principles allow one to modify in the understanding of those principles.

The characteristics of mixing and of lay down of the mixed chemicals are of major significance in the control of the wide variety or urethane foams now available for laminations to textiles. In particular the range of cell size, colours and densities involved are under constant study.

Nozzle design is particularly significant because of the effects of dead spots in creating pre-nucleation and bubble contamination in the rising foam. The effects of impurities and how one can remove impurities or counteract any effect that impurities may have are under constant review.

SLITTING VS PEELING

Since in slitting one always cuts across the surface of the block at right angles to the direction of rise one gets, providing the block is free from striations and so on, and even mattness of surface. In peeled foam, because during the course of peeling one cuts through the block top, bottom and down sides (along the direction of rise) one sees a variation in cell orientation which shows up as bands. This change in cell orientation due to peeling can affect foam properties and the foamer must offset this, as far as possible for formulation and cell size control.

ROUND BLOCK VS SQUARE BLOCK

Yields can be obtained, in the case of a block section, peeling at say 0.75 mm, of over a thousand yards per block which means that for the average call off of a particular colour it is quite possible at this particular dimension, to keep in stock a sufficient quantity of blocks to accommodate the small order in the 1000–3000 yards range.

FOAM TYPES RANGE

I will now summarize the types of foam available for laminating particularly in the flame lamination field. In say 1961, we were restricted to greys, beige and white foams; very little in the field of coloured foams was available nor was it in fact considered really necessary. With the development of "mini bond" foam lamination and the use of foam primarily as an adhesive, insulating and stabilizing layer between somewhat unstable fabrics, colour match foams were considered necessary. The range of foam colours available now is very extensive from most suppliers with as many as 30 to 40 different shades available. From the original marks of 2–2.5 mm or thereabouts we now have foam lengths available up to a thousand yards or more with thicknesses ranging down to 0.5 mm.

LOOKING FORWARD

If one were to look forward, the areas for development that would offer the most interest in the light of the way the process is developing at present would be (a) progressive improvements in foam production and (b) better understanding of the parameters of lamination to various fabrics and fabric types, dyeing etc., and lastly but by no means least the development of the raw material suppliers of new additives to follow the improvements demanded by the better understanding of the items (a) and (b).

B. T. WATERS
General Manager of Development

E. J. Arnold & Son Limited
Butterley Street
Leeds LS10 1AX England

A RETAILER'S VIEW OF THE USE OF POLYURETHANES IN THE FASHION TEXTILE INDUSTRY

(Received January 7, 1974)

INTRODUCTION

AT ONE TIME perhaps "fashion" referred only to the clothes that a small proportion of the female population wore. In most other fields changes in colour, materials and style could only be observed in decades rather than seasons. This is not nowadays true, and increasing competition and increasingly effective communications media have on the one hand demanded and on the other enabled the consumer to be presented with an ever-changing range of goods. Certainly all garments, men's, women's and children's – other than those of a purely functional or protective type – have now an element of fashion about them. That footwear and accessories are subject to similar pressures seems at first sight a truism – but remember I include men's and children's as well – so it's a truism of our times perhaps. The same pressures are now beginning to come to bear on luggage and furniture – perhaps only the future will tell us just how much effect fashion swings will have on this market. If we consider textiles in the broadest sense, we must include "non-wovens" as well as the weaves and the knits. Once we consider these then the field opens very wide – perhaps we might include the PU coated papers we have seen recently.

We have then three main areas of application – the PU foam backed textiles, the PU coated textiles and the PU poromerics. In addition, we shouldn't lose sight of the PU elastic fibres giving fabrics of the Lycra type. These are, of course, largely confined to corsetry, but are beginning to find applications in swimwear and footwear.

BACKGROUND

The PU foams were introduced to change the function of cheap light fabrics from dress materials to coat materials. Initially the biggest technical problem was delamination during wear or subsequent washing or cleaning. There was of course, also the aesthetic problem of turning everyone who wore them into "Michelin

Men". As the technology grows more sophisticated, there is perhaps a tendency for foam laminating to be carried out to alter aesthetics rather than function. Some of the "mini" – foam laminates now available are of this type.

The PU coated materials on the other hand, were introduced to change aesthetics for coated fabrics – oil, nitrocellulose and PVC coated – have been with us for many years. However, PU offered a lightness of handle and a level of gloss not possible with conventional materials. Its other characteristics enabled PU materials to be used for other functions including one of the few innovatory uses of recent times – gaiters.

The PU poromerics – where the aim is to imitate leather by imparting a porosity – generally have a non-woven, a woven or a knitted substrate, or a combination of non-woven and knit or non-woven and weave. In these cases the textile is subsidiary to the main purpose, and is there generally to reinforce the PU. In this case the textile upgrades the PU – not the PU upgrading the textile.

SPECIFICATIONS

What view of PUs has the consumer? As "PU", probably none, but as dry-cleanable, wet-look, as windproof, light summer fabrics, or as elasticated garments that won't tire after washing, they will have quite a lot. In other words, the consumer is not interested in what it is – only what it looks like and how it performs. The consumer will, of course, be critical if the items fails to live up to its initial promise, and particularly critical if performance of replaced or repeated items is similarly suspect. In as much as the retailer must try to give the consumer what he wants – or what he thinks he wants – the retailer will mirror this approach – price and performance (including aesthetics) will be the main points.

In addition, the retailer has other pressures. The consumer has perhaps a year or more to assess performance. A retailer may only have a few days. Littlewoods – through their stores and catalogues sell about 6000 items each season. The cycle of the seasons demands a bi-annual change, and since from conception to death is a brief 18 months, at any one time some 20,000 items are under consideration. Fashion may require a particularly quick decision – the gaiters I have already mentioned took only about six weeks from first being seen to being on sale. Under these circumstances, the retailer's view of a material is bound to be superficial and his specifications "fail-safe" – he must quickly reject any materials that are at all suspect, and if this occasions the rejection of materials that might just be satisfactory and the acceptance of those with belt and braces safety, that's all part of the game. Progressive fining down of specifications to an acceptable level can be carried out in an evolutionary way in successive seasons.

Normally in drawing up a specification one has some idea of the sort of parameter that one is going to choose from other materials. In considering PU coated materials, we perhaps would consider date from pure textiles, other coated fabrics and leather in their various end use applications in the first stages. This would lead us to decide to specify tensile strength, tear strength, burst strength, abrasion, flexing and colour fastness to light etc. At this stage we aren't sure what levels of these various parameters we will choose, we may have to carry out associated wear and laboratory test programmes. Our tendency is to keep the tests down very much to a "short list" – a proliferation of highly academic tests, although impressive on paper may add but marginally to our state of knowledge of a particular material.

The next stage is to consider the demands of specific end uses one against the other. Thus garments must be cleaned – one way or the other. Shoes must be lasted and so on. These considerations enable us to decide which special tests that each end use material must undergo.

Finally, we come to that part of specifications that is germane to the particular garment or item of the end use. This cannot in general be specified in advance, and is left to the garment technologist, rather than the cloth or textile technologist to decide. It concerns the style of the garment – what will the stresses be in wear, what sort of seaming will be used (particularly in decorative panels), what sort of trims or motifs will be used and so on. In this case each line must be looked at individually and the components checked for compatability.

By working in this way we can in fact, build up a sort of identikit specification, and a set of such specifications is shown in the table. Having built up such a kit, we can use the components of it to draw up a new specification. Suppose we take as an example "hot pants". Recently we have been looking at wet-look hot pants. Hot pants are a garment and therefore will have to be washed or dry cleaned. On the other hand, their aesthetic appeal is in their stretch properties. For these foregoing properties we can use our identikit, choosing wash and fastness properties from the garment specification and physical properties from the boot-legs and gaiters specification. But what are the special demands of hot-pants? The original garment specification was conceived for fairly loose fitting outer garments, and the dye fastness of the backer to perspiration and urine stains were not considered of critical importance. In hot pants it will be. Again, will the presence of perspiration cause hydrolytic damage to the PU coating or the bond between backer and PU coating? Perhaps after some discussion we might conclude that a material for hot pants should conform to the following minimum standards.

Table 1.

Tensile strength		40 kg/5 cm	
Tear		2.5 kg	
Abrasion		100	
Coating adhesion		0.5	
Flexing		100 000 (BS 3424, 11B)	
		Shade	*Stain*
Colour fastness–	light	4	4
	wash	4	4
	dry clean	–	–
	wet rub	–	4 (back only)
	dry rub	–	3 (back only)
Shrinkage	wash	2%	
	dry clean	1%	
Fryma		50% if stretch at 3 kg/5 cm load	

In addition to testing against the specifications, there are additional tests that have to be carried out. Any special claims as to washability, durability etc. must be checked. In this context I have been particularly disappointed recently in trying to check claims made by certain Continental companies for PU based suede substitutes. Claims for performance, particularly washability and stain and spot resistance have been difficult to substantiate, and no systematic testing seems to have been carried out by the manufacturer. In the UK – with our Trade's Descriptions Act – such claims must not be made unless they can be substantiated. PU imitation suedes are not cheap – in many cases they are as expensive as the genuine article, and they will only sell against cheaper and less robust substitutes and indeed against suede on their special claims.

Table 2. Suggested Specifications for Smooth PU Coated Fabrics

	End use		Shoe uppers	Boot leg, Gaiters	Accessories	Gloves	Garments
	Special properties required		Lastability scuff/abrasion resistance	Stretchability		Hand washability	Washability and durability in wear
	Weight (not specified)		10 oz./sq. yd approx.	5.6 oz./sq. yd range	4 oz./sq. yd approx.		-
Specification, minimum standards	Tensile		40 kg/5 cm	40 kg/5 cm	30 kg/5 cm	12 kg/5 cm	40 kg/5 cm
	Tear (single rip)		2.5 kg	2.5 kg	2.5 kg	2 kg	-
	Burst		--		--	-	10 kg/cm²
	Abrasion (Martindale)	28 oz. worsted	--	--	-	-	50 000
		21 oz. 600 grade emery	600	400	200	200	400
	Coating adhesion		1.0 kg/cm	0.5 kg/cm	0.4 kg/cm	0.4 kg/cm	1.0 kg/cm
	Flexing		1 000 000 (Satra vamp without grade A cracks)	75 000 (Satra tension flexer)	-	50 000 BS 3424, 11B	100 000 BS 3424, 11B
	Colour fastness	Light	4 (face only)	4 (face only)	4 (face only)	4 (face only)	4 (face and back)
		Wash ISO 1 + Daz	--	-	–	4 (shade only)	*Shade* 4 *Stain* 4
		Dry clean	--	-	–	--	4 4
		Wet rub			*Shade* - *Stain* 4 (face only)	--	-- 4 (face and back)
		Dry rub		–	-- 3 (face only)		- 3 (face and back)
		Water	–	*Shade* 4 *Stain* 3	4 3		4 3
	Other tests		7 mm Lastometer	Flyma 50% stretch at 3 kg/5 cm load		Wash shrinkage 2%	Wash shrinkage 2% Dry clean shrinkage 1% Blocking (3424, 13) 100 g weight no sticking or damage

What improvements must be effected in today's PUs? In the best little or no improvement is necessary. However, the consistency of materials needs to be improved – not merely from manufacturer to manufacturer, but from batch to batch within manufacturers. We would like to see then a general lifting up of performance standards.

FUTURE CONSIDERATIONS

What does the future hold for PUs? Immediately I think that fashion predictions home strongly towards suede, and in this context synthetic materials are merely imitative. It is perhaps possible to perceive that synthetics could be used to give "suede and nappa" patterns by a combination of heat, pressure and chemical treatment of the surface – effects as far as I know impossible with skins. Thereafter fashion is in the lap of the gods, and perhaps in the lap of the development technologist now working on tomorrow's equivalent of the "wet-look" – a genuine innovatory effect – in today's laboratory.

This paper originally presented at the Urethanes in Textiles Symposium, Manchester, England, May 1971, sponsored by The Institution of The Rubber Industry, London.

TSU HUAI WANG AND ROBERT B. ORR

Wilmington Chemical Corporation
Wilmington, Delaware 19899

EARLY FINDINGS RELATED TO THE USE OF SOLVENT-FREE REACTIVE POLYURETHANE SYSTEMS FOR COATED FABRICS

(Received September 18, 1974)

ABSTRACT: This article describes preliminary work carried out to develop a solvent-free reactive foamed urethane coating for fabrics. Such composites of non-cellular and cellular surfaces will most probably lead to a wide variety of new products.

KEY WORDS: Urethane composites, flexible coatings, fabric coating processes.

INTRODUCTION

OVER THE LAST half century, the treatment of fibrous substrates has seen an evolution which began with the filling of fabrics with natural products and has progressed to the surfacing of a number of substrates with many types of coatings. The development of new coatings materials and new coatings techniques has more completely filled the demands of old applications and has generated new applications based on the remarkable physical and aesthetic qualities now obtained. We can get a quick appreciation of this evolution by giving a glance at earlier examples, such as waxed cotton canvas and rosin filled linen, and a close inspection of today's coated fabrics, some of which are foamed vinyl or are transfer-coated urethane composites. This paper will summarize the results of our initial studies using a solvent-free reactive system in a transfer coating mode. A brief commentary will be made as to the further developments to be sought.

THE COATING PROCESS

Chemical developments have provided durable, flexible coatings, while application techniques have been engineered so as to allow handling of a broader range of substrates as well as new coatings materials. A summary of many of the processes in current uses is in Figure 1. In this list are shown two categories of processes. With the methods using thermoplastic material, well engineered heavy duty equipment is required. With the processes using solvent-based thermoplastics or reactive systems, the equipment may be of a lighter weight design. This difference provides for a

Figure 1. FABRIC COATING PROCESSES

Generally Thermoplastic	Melt extrusion of film(s) – laminating of melt(s) directly or adhesive laminating to substrates
	Calendering of film(s) – followed by laminating
Can be reactive, crosslinking systems	Knife coating directly onto substrate Printing of coating onto substrates Spray coating Flow coating Transfer coating

lower initial investment for the latter system. Product shown in Figure 2 is from a transfer coating process. In this particular instance, the substrate is a woven, napped fabric. Transfer coating has two major advantages over some of the other procedures. The texture of the applied surface can be varied by the selection of the transporting medium and the penetration of the surfacing into the fabric can be very closely controlled. In order to give the basis for a better understanding of this work, a brief description of the current transfer coating procedure is given.

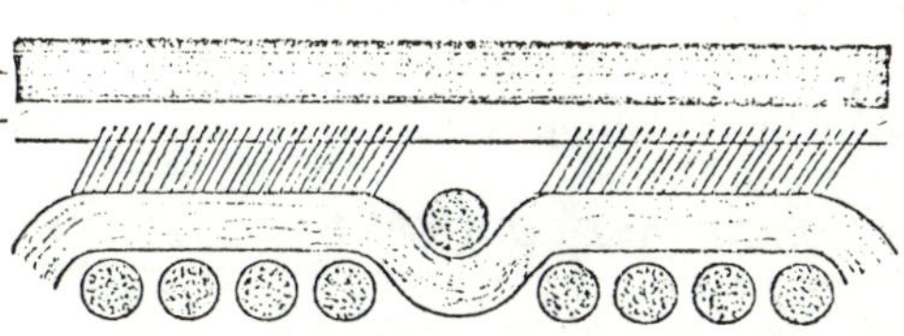

*Figure 2. Idealized Structure of Polyurethane Coated Fabric**

The top coating is applied to a transporting paper which will allow stripping of the film subsequent to drying or curing. This film is dried and left on the transport paper. An adhesive layer is then applied and the substrate, in the case shown, woven fabric, is then brought into contact with the adhesive. The adhesive is dried or cured and the composite, as illustrated, is stripped from the paper. With many of the urethane products in use, these composites have a total thickness of coating of 50 to 80 microns. Such processes can readily operate at lineal speeds of 10 to 15 meters per minute.

In Figure 3, there is shown a composite made up of the woven fabric, a microcellular surface, and a skin coat again made by a transfer coating process. The transfer coating procedure gives the opportunity to make such structures having many and diverse elements in the final composite.

Some of the factors which we feel are of value as demands on long range development efforts are listed in Figure 4. We have used this list as a guide in our program to investigate new chemical and mechanical systems for coating fabric. Our

*D. Popplewell and L. G. Hole in "Urethanes in Coated Fabrics," Technomic Publishing Co., Westport, Conn. (1973).

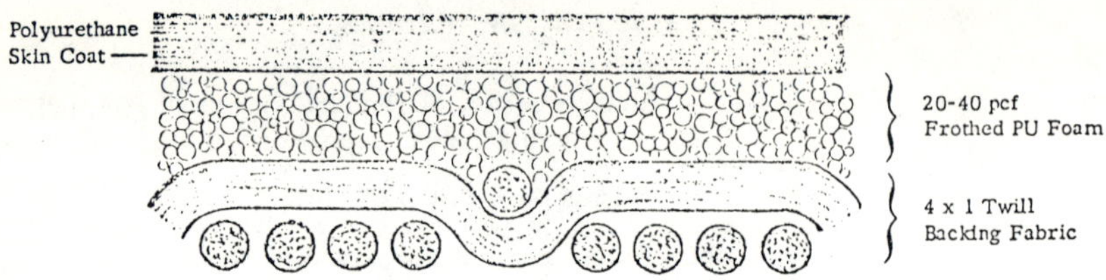

Figure 3. Idealized Structure of Polyurethane/Fabric Composite Having a Frothed Urethane Core

work has followed this path. Many types of urethane systems are available as shown in Figure 5 [1]. Of these systems, the two-package approach holds the promise of lowest solvent and energy requirements. An obvious existing example of this is urethane foam. The solvent contained in a urethane foam generally fills two functions. It serves to reduce the viscosities of the streams and provides a source of inert gas which is a factor in the final density of the product. The existence of non-cellular cast urethane elastomers gave evidence that such an approach would also have promise for the production of a non-cellular surface. Thus, it appeared that a means to a coating with greatly reduced solvent requirements was feasible. Urethane chemistry with its proved versatility was combined with urethane foam mixing and materials handling technology.

FIGURE 4. LONG RANGE GOALS OF FABRIC SURFACING EFFORT

Process:	Minimum Energy Requirements Minimum Personnel Hazard Non-Polluting
Product:	Minimum Hazardous residual reactants or additives Non-Flammable or Combustion products non-toxic Recyclicable and/or biodegradable Broad range of physical properties

FORMULATIONS

While there are many possible formulations with which we could have started, we selected an approach from the patent literature [2]. This system appeared to have a balance of properties holding promise as a coating for a flexible substrate, such as fabric. Figure 6 gives the formulation used as well as physical properties. This system has been frothed on foam equipment to make the microcellular composites pictured in Figure 3. This structure gave an improved hiding of the substrate texture, which is of some commercial promise. These composites of non-cellular and cellular surfaces with a fibrous supporting web will, it is believed, lead to a wide variety of interesting products.

FIGURE 5. TYPES OF POLYURETHANE SYSTEMS

Non-Aqueous, one-package systems
- Oxidative Cure
- Moisture Cure
- Blocked Isocyanate Cure
- Amine-Formaldehyde Cure
- Thermoplastic
- UV Cure

Non-Aqueous, two-package systems
- Isocyanate Terminated Resin Plus Catalyst
- Isocyanate Terminated Resin, Plus Polyol,
Plus Polyamine, or
Plus Ketimine

Aqueous System
- Nonionic Solutions of Thermoplastic
- Emulsified Urethane Lacquers
- Ionic

High Solids Coatings

FIGURE 6. STARTING FORMULATION AND PROPERTIES OF FROTHED FOIL

Niax Polyol 31–45	74 pbw		
Niax Polyol D510	14 pbw		
Dipropylene Glycol	13 pbw		
TDI/D510 Prepolymer	50 pbw		
Reaction Regulants			
Density, g/cc	0.16	0.4	0.5
Tensile Strength, kg/sq cm^2	10	20	39
Elongation to Break, %	300	290	310
Tear, g/cm	5.2	8.6	10.
Cell Size Range, Micron	40–250	40–310	40–360

The author's company has produced coated products utilizing less than 3,000 molecular weight – raw materials, no solvent, and the heat of reaction to assist cure. This was done on a transfer coating machine. The mixing and dispensing equipment was obtained from foam applications. The development of the technology to a state required to make the concepts more completely and broadly functional should be undertaken. These efforts should include the development of means to control cell size and cell uniformity as well as the physical properties. In our studies, cell size ranges have been from 40 to 360 microns. A successful development will likely require a close coordination of the chemistry and of the mechanics used: A close coordination of the chemistry of the coating (selection of polymer types and specific polymers of the particular types), the mixing, transport and application technique, the interaction of the chemistry with these mechanics and the specific substrate used. We feel that the architecture of these products holds great promise. It is expected that this route, which is to carry out the polymer synthesis in situ, will lead to a wide range of properties and may extend beyond urethanes to include other polymer systems.

ACKNOWLEDGMENTS

The Authors wish to acknowledge the assistance of Union Carbide Corporation in technical assistance and Norwood Industries which was instrumental in obtaining evidence of the practicality of this chemical/mechanical approach to surfacing.

REFERENCES

1. W. Chang, R. L. Scriven, J. R. Peffer, and S. Porter, Jr., "Advances in Polyurethane Coatings (1969 to early 1972)", *Industrial Engineering Chemical Products Research Development,* 12, No. 4, Pgs. 278–288, 1973.
2. Lawrence Marlin and Eric Schwarz, *British Patent Specification, 1306372*, February, 1973.

LAWRENCE B. MOSS

Witco Chemical Co.
New York, NY

CURRENT AND FUTURE APPLICATIONS FOR URETHANE COATINGS ON TEXTILES*

(Received October 15, 1974)

SINCE THE 1960'S urethanes have been used to coat both non-porous as well as porous substrates in the United States. The non-porous substrates would, include such applications as pigmented enamels, seamless flooring systems, metal finishes, wire coatings, rubber finishes, etc. However, the subject of this paper includes porous substrates and in particular; current and future applications for urethanes coated on textiles.

Utilizing urethane chemicals for textile coatings has realized an impressive growth over the past five years. However, in most applications, their greatest noticeable affect has been made in the consumer market.

This is where urethane coated materials will have their greatest impact for growth, or their most striking reason for failure.

The markets that will be discussed include apparel, accessories, upholstery, footwear, some limited industrial applications, and what many feel to be the largest potential for urethane coated fabrics, automotive applications.

The fabric coatings industry has achieved some degree of expertise in the areas where minimum standards are required. These areas would include the apparel, handbags, gloves, and personal "leather" goods markets. Here the stylists and the coaters ability to interpret their ideas and achieve asthetics along with minimal quality standards will provide an acceptable product. In the areas of higher performance, markets such as upholstery and shoe uppers the market has not achieved the best products possible.

In the market areas of luggage and automotive the United States urethane coaters have made very little progress to produce a commercial product.

Industrial coatings on textile substrates include another facet and potential for continued growth area for urethanes. Here the requirements, although stringent, are quite specific and aesthetics usually play less of a determining role in the acceptability of the product.

As each area is covered, no attempt will be made to project yardage or specific growth potential for these market areas. Rather, the future for growth will be

*Presented originally at Urethanes in Coatings Seminar, sponsored by the Program Division, Technomic Publishing Co., Inc., Westport, CT 06880.

dependent on several factors – some of which are intangible and frankly, unpredictable. However, the other factors of styling, quality, availability, and consumer acceptance will play a major role.

APPAREL

The apparel industry is one of our country's largest. It also has the dubious honor of having a high degree of risk due to the rapid style changes that are the stock and trade of the industry.

Urethane coated fabrics have enjoyed a good growth in this market because of their ability to be readily styled and modified to fit seasonal and designer changes.

Leather has always been a relatively stable portion of this market and its appeal holds season to season. A style of leather, suede, is also popular traditionally. However, leather is expensive and must be specially cared for.

Urethane coated fabrics for apparel became popular in the United States in 1969 with the advent of the wet, crinkle look in women's coats and rainwear.

However, because of delivery pressures for large yardage during that critical season the poor quality became evident in consumers returning coats that literally fell apart. Hopefully the coaters learned from that costly experience and have demanded the best product from their supplier. The basic criteria for an acceptable urethane coated fabric for apparel include drycleanability, washability, scuff resistance, softness and pliability in cold weather, drapability, dry hand, good adhesion of the skin coat to the fabric, a leather look, and light weight.

The past five (5) years have seen many variations and the next five (5) years will continue to see these variations on a theme.

Suede has become a staple item in the Fall 1974 market and it may be achieved using urethane coated fabrics via direct coating and brushing, flocking over a urethane coated fabric, via impregnation of a substrate and brushing, or by utilizing a urethane microporous foam system over fabric and abrading the surface (a standard method of making a PVC suede).

Over the past two (2) years there have been no major breakthroughs in urethane coated fabric styling for apparel. (The related markets of handbag, gloves and personal leather goods seem to follow the general trend.) Variations are appearing this Fall as in previous seasons but because of the general economic situation business has not grown as dramatically and the changes have been subtle.

The related industries which would include handbags, gloves, and personal leather goods do not seem to be viable enough to stand on their own. It seems that aesthetics and price are the determining factors and almost any thermoplastic that will do the job is sufficient.

The three other areas of footwear, upholstery and automotive applications require higher performance and their future will be determined in large part as to how well they perform initially and how they will age.

Because each of these areas can be considered in its infancy and their growth is so dependent on the quality of the coated textile they should be evaluated separately.

FOOTWEAR

The market that has had the most dramatic growth in the past few years can be found in footwear. Several factors can be attributed to this and they include the general shortage of leather, advanced styling, and it has evolved that some of these snythetics will outperform leather. Price wise, urethane coated fabrics for footwear fit between the higher priced leather and poromerics and the less expensive vinyls. Before discussing the shoe upper business and its future it is important to note that about five years ago the first wide use of thermoplastic urethane for footwear made its appearance in the form of womens' boots. The ability of the urethane to stretch and conform to a woman's leg, the comfort, and the high styling possible made their use highly desirable.

The past two years has not seen as much business for the boot trade, primarily because of style changes. However, as the thermoplastic urethanes continue to grow the coated fabrics will find their way into more boot production.

This will happen primarily because the growth for urethane in shoe uppers will continue. There are new customers, higher product acceptance, more styling and new developments in the chemicals and the textile substrates being used.

The use of urethane coated fabrics followed close behind the use of poromerics. Besides the actual process of making a poromeric the major difference between the two is the fact that poromerics are produced on a non-woven. owever, this is changing since there are transfer coated fabrics being made on non-wovens and this will continue to grow.

A compilation of market studies indicate that women's shoes account for almost one-half of the total shoe production in the United States. Womens high style shoes are where urethanes have had their greatest impact to date in the footwear trade. Resin producers have come forward with a host of high quality products which are available to the coater.

These skin and tie coats provide the construction with better film to fabric adhesion and better flex life which provides the shoe with resistance to delamination, cracking, peeling and abrasion in wear. The problem of hydrolytic stability in these shoe upper materials has been of concern and it is associated with the chemistry of the urethane product. However, this problem was most prevalent in Europe where thermoplastics were not used to a great degree at one time.

In general, the future for urethane in shoe materials seems bright. Mens shoes offer an excellent potential and products have been offered to the coaters which appear to me meet the higher quality criteria. Children's shoes present a problem of having to offer a high quality product in a lower priced market. Competition with vinyl will be difficult.

One of the more dramatic changes that will occur can be found in the substrate. The textile substrate being used for the footwear market plays an important role in the final quality of the good produced.

In general, the quality of domestically produced sateens is not as good as those being imported and new substrates are being considered by some. Included here would be polyester/cotton blends, synthetics, and perhaps the most important of all in the long run non-wovens. Non-wovens with proper characteristics are currently being used successfully in Italy. A possible approach for a non-woven will be discussed later in the paper.

Woven sateens tend to unravel during the curring in the manufacture of shoes. The use of a non-woven substrate will eliminate this problem for the shoe manufacturer.

UPHOLSTERY

Although upholstery had a fairly good start in the United States in 1968 it was soon apparent that the quality was not good enough to withstand the expected life of a piece of high priced furniture. The light weight used was really an apparel grade and there were considerable complaints from the consumer which included delamination, cracking, poor color stability and a host of others. Suits that grew out of these initial attempts at upholstery are still being arbitrated.

However, even with failures that were occurring, literally around the world, there was enough interest in the new material to support further work to develop acceptable products.

The initial products gave the designers a new material to work with an; it processed into an apparently attractive alternative to vinyl with many of the positive qualities of leather.

Time has past since those initial attempts and it appears that the furniture trade is again willing to seriously look at urethanes. True, there has been urethane coated fabrics offered on the market all along and their quality has improved but the real test will occur in the next few years as the major furniture houses being their programs once again.

It is up to the urethane supplier to educate the coater and the furniture trade of what kind of product will give the best aesthetics and provide the quality that will make the business available a viable part of the industry.

With these criteria being met the future for upholstery will lie in household furniture (excluding kitchen sets for the most part), re-upholstery, decorator trade, and other areas where the price of urethanes will not weigh against their use.

Because of the price structure for urethane and the need for "armour" rather than high style, upholstery for use in commercial, public buildings, office furniture, kitchen furniture, etc., will be limited. In these markets price and performance count rather than any styling appeal urethanes may have. Urethane topfinishes may provide this sector of the upholstery trade with a protective finish that will improve the abrasion resistance of the plastic materials used and also help protect the printing used in their manufacture. There are urethane products being offered on the market for upholstery that will provide light stable films with a construction having good abrasion resistance, flexibility and recovery, and outstanding hydrolytic stability. It is also felt that these products will provide a good degree of resistance from what many feel is the major reason for the initial failures of upholstery fabrics made from urethane-microbial decomposition.

Along with these recent developments there are areas that furniture houses would like to see improved. They include spot and soil resistance, improved durability (always looking for), and non-flammability.

There are more and more Federal and State laws covering the use of home furnishing materials with regard to flammability. It would behoove resin suppliers to include flame resistance as part of their development program before a regulation is written which would have a major effect.

The substrate used in coated fabrics for upholstery will probably continue more towards the knits. Knits have the advantage in upholstery of stretching and conforming to the shape of the piece which would have definite advantages in ease of handling.

AUTOMOTIVE

The automotive industry is the largest single user of coated fabrics and for that matter a major consumer of plastics of all types.

The average car uses approximately 4–5 linear yards of fabric for upholstery. Thus a U.S. production of 15,000,000 cars each year would represent sixty to seventy-five million linear yards of upholstery fabric. Of this approximately 70% use vinyl coated fabrics today.

As the development program by both resin suppliers, coaters, and potential users progresses it is obvious to see that there is an excellent market for urethane coated fabrics. Of course the requirements for a suitable material will be unlike any being sold in the marketplace today.

In the early days of the automobile the upholstery was either leather or woven fabric. Leather was used for open cars and as is true today for luxury styling. The first vinyl coated fabrics were introduced in the mid-forties and today they represent about 70% of the upholstery materials being used.

The major drawback of vinyl fabrics has been one of comfort. They get very hot in the summer months and very cold and stiff in the winter. Designers have attempted to style around this by using combinations of vinyl and fabric.

The heat resistance for an automotive application is an important consideration, since the upholstery is exposed to temperatures ranging from minus 30 to 200 degrees Fahrenheit.

In the case of nearly all materials used in an automobile the two principal features are economy and performance. This is the reason vinyl has enjoyed such a growth in the automobile industry and anything new to come along will at some point be compared to vinyl. However, it would be most unfortunate if the automotive engineering and research departments did not take into consideration that vinyl and urethane differ in areas for very practical reasons. For instance it is generally accepted that a vinyl coating weight of approximately 15–18 oz per square yard is necessary while a urethane coated fabric will perform as well at much lighter weights. This difference alone will help in the cost factor of urethanes.

Every major automotive company seem to be ready to use urethane coated fabrics but to date the engineering departments have not approved the performance of those products evaluated. Various market reports have stated that urethane would be used in the 1975 model year, but its use will probably be limited to the higher priced, higher styled models and primarily on an experimental basis.

It is important to note the advantages and disadvantages of urethane coated fabrics so their feasibility can be better understood.

The advantages may be listed as follows:

1. leather look with a dry hand
2. no odor or toxilogical problems due to plasticizer migration
3. provides a soft fabric with great flexibility for tailoring.
4. comfortable cool with better moisture vapor transmission resulting in comfort both in summer and winter.

The disadvantages to date would have to include:

1. no product has been approved for abrasion and scuff resistance and yet provide a soft flexible hand. Detroit requires a piece of goods that will hold up for at least 5 years.
2. flammability standards are continually being upgraded and the products used must meet these ever demanding specifications.
3. The upholstery must withstand prolonged periods of heat, cold dryness, or moisture as well as extreme changes in any of these conditions.
4. Economics are important if urethane is to make any inroad in the vinyl currently being used. Urethanes appear to be competitive, with the woven fabrics and leather.

This market certainly gives the urethane producer and coater something to work towards over the next year or so. However, past mistakes and experiences should certainly make each as cautious as the automotive people will be in evaluating their products. A black eye in this market can potentially touch every consumer and do irrepairable harm.

In covering the basic markets for transfer coated fabrics it is important to mention direct coating, especially on nylon. Direct coating techniques are primarily used in textile applications or the industrial market. During the Viet Nam conflict many yards of urethane direct coated nylon were being produced for such applications as ponchos, tenting, rainwear, tarpaulins, etc. Now that the demand has dropped for this material the direct coaters are looking elsewhere to expand their business. Much of the yardage they are producing is going into leisure areas such as golf jackets, tenting, skiwear, outerwear and the like. Also skimobile tarps, tarpaulins for commercial use, airplane emergency chutes and inflatible life rafts.

Growth for this area would seem to be in the improvement of the urethanes being used. Much of the urethane being used is a one package aromatic with a crosslinker built. Non flammability and better heat sealibility would be an immediate improvement in the products being used. Also, improved hand and non-curling qualities would be desirable.

As far as other materials for applications in urethane coatings and adhesives in textile applications, there are two areas that are coming on strong . . . water based urethanes and 100% solids urethanes.

With recent air pollution laws and PSHA requirements being enacted, the need for these types of materials is growing and many of the urethane suppliers have active programs along these lines.

Water based urethane materials have been around for some time. Because of their cost and the fact that early products did not work as well as other types of latex on the market, they did not fare too well.

However, the time has come and there are several materials being offered. Many have been compounded to various degrees but there are also latices being sold uncompounded with relatively simple thickening and compounding techniques outlines. The immediate potential use for these materials is an adhesive tie coat for transfer coated fabrics. However, there are a host of other textile applications where the inherent qualities of urethane materials make them most desirable. These would include fabric backcoating, fiberglass yarn coating, etc. Perhaps one of the greatest potentials lie in the manufacture of non-wovens.

Because of urethanes inherent properties, they can offer a non-woven abrasion resistance, flexibility softness, or firmness, possible use of cheaper fibers, improved light stability, etc.

It should be noted here that the urethane latex materials being offered do not require additional heat or catalyst to achieve their full properties whereas many of the other materials available or being used need this to pass such tests as wash and dry cleaning tests.

The urethane latex can be mechanically frothed to provide a strong foam which may be used as an additional tool for the coater to work with the stylist.

The manufacturers of these urethane latices are continuing their programs to improve the products and although some are being recommended for use as a skin coat their wide use along these lines must be improved upon.

With water based urethanes the greatest problem in handling is to remove all of the water during curing. The ability and cost of doing this could lead into the next area where much work is being done . . . 100% solids materials. This presents a solvent free, 100% solids material that could typically be used via roll coating utilizing such equipment as the Zimmer or Maurer machines. An alternative would be to use present equipment with some modification.

These new materials are being examined by many in the United States and the interest is growing every day.

There are many questions being asked and the answers are not clear in many cases. What pigment systems can be used? Will the product give the necessary physical properties without loss of aesthetics? Will special metering heads be required? Much time and money is being spent in research and development to answer these questions and alleviate some of the intricacies of using a 100% solids urethane coating system.

Another tool the coater might find of use in the styling of products for the future is microporous foam. This type of material is unique in that a urethane foam can be produced in a thin layer, in line, on present coating equipment. No extraordinary mixing or handling is needed and the use for this material can be varied depending on the requirements outlined.

The continued growth of these varied markets and ideas will depend to large degree on the ability of the basic chemical producer to continue his development programs and work with the coater and marketplace.

C. BLUESTEIN, P. LOEWRIGKEIT T. T. MCGIMPSEY,
AND K. A. VAN DYK

Witco Chemical Corporation
100 Bauer Drive
Oakland, NJ 07436

FACTORS INFLUENCING LATEX TIE-COAT ADHESION IN URETHANE SKIN COAT TO FABRIC BONDING*

(Received September 26, 1974)

ABSTRACT: The factors affecting adhesive performance in a coated fabric construction are considered, particularly with respect to bonding of cotton to polyurethane using a polyurethane latex tie-coat. An adhesive peel strength test method is detailed for hand evaluation. Improvements in reproducibility of results have been obtained using a template constructed from Mylar film mounted in an aluminum frame in order to prepare peel strips. Precision was improved, but due to inherent variability in hand application techniques it was not optimized. Relative effects of a number of factors affecting peel strengths in the systems under study are discussed.

KEY WORDS: Urethane coated fabrics, urethane latex, peel strength, tie-coat, adhesion, urethane skin coat, template, transfer coating.

INTRODUCTION

ALTHOUGH COATED FABRICS technology is no longer novel, current developments in the industry will lead to new and wider ranging products and applications for them. Newer resins and formulations for this technology are becoming available. Some of the factors contributing to the development of the newer materials are the shortages of natural fibers and skins, and also of common coatings materials, and the need for improvements in processing efficiency with respect to worker and environmental hazards and the need for more efficient utilization of energy sources.

Our current R&D program for improved coatings polymers is concentrated on the polyurethane basic materials. During the development of water-based products, in particular, it was desirable to develop preliminary laboratory bench screening tests in order to evaluate product suitability and reproducibility. There are many such empirical evaluations in common use, but we have found that any consensus that we have tried to compile indicates that similar tests are run by sufficiently

***This paper was originally presented at the Symposium on Coated Fabrics, American Chemical Society, Atlantic City Meeting, September 1974.**

modified procedures that it leads to differences of results, and therefore to disagreement between laboratories when an attempt is made to correlate results. The problem has been widely recognized, and organizations such as AATCC in the United States and SATRA in Great Britain [1] have active programs underway to assemble and standardize test methods for coated fabrics technology. Also available are methods published by the British Standards Institution [2] and the Chemical Fabrics and Film Association [3]. These methods assume that one has a sample of coated fabric available for testing, presumably the product of a coating machine.

A most desirable objective would be to have a hand method which could, without elaborate manipulations, be correlatable to machine coating and be reproducible from laboratory to laboratory. The procedures presented below represent some advancement toward that end.

One of the important considerations in development of new materials for coated fabrics is how they contribute to adhesion of the completed construction. Peel tests are the accepted procedure to evaluate the adhesive performance, but within the nominal standards set, the variants and interpretations are as each laboratory prefers to operate. Thus literature reports usually state or imply "relative peel" [4, 5, 6, 7]. In our laboratory we have used the common and convenient means of determining pounds of force to peel a 1-inch wide strip in a T-peel type test. In this paper we will discuss some of the parameters one needs to consider when using water-based polyurethanes to prepare fabric constructions for peel tests and suggest some techniques for improved adhesion.

LABORATORY PREPARATION OF COATED FABRIC COMPOSITES

In a simplified concept a urethane coated fabric is made of three parts: a fabric backing, a tie-coat for bonding, and a surface skin. Since the majority of machine coating today is carried out by the transfer process, the laboratory technique is based on the same principles.

A urethane skin coat is cast on silicone treated release paper. A paper which we have found satisfactory is S. D. Warren Transkote Pat. AV-CIS. A convenient paper width is 10-inches. It can be weighted or taped in place on a smooth, flat bench. The skin coat, which can be clear or formulated, solvent based or as aqueous dispersion, is made up at about 15,000 cps. This coating is spread with a Gardner knife set such that it can produce a film which when dry is 2–3 mils in thickness. Alternately an appropriate sized RDS rod (Maier bar) can be used for spreading. The cast film is air dried for 15 minutes then oven dried 5–10 minutes at 110°C.

The dried skin is cooled and should be uniform, free of flaws and still adhering to the release paper. A tie coat is placed on this skin by drawing the latex down with a No. 10 Maier bar (RDS rod). Immediately the laminate is completed by smoothing down upon the surface a piece of fabric which has been wound on a hand held roller. The completed laminate is then placed in an oven at 110°C for 5–10 minutes after which it is dried and conditioned to test temperature. In order to prepare peel sample constructions, two of the laminates are glued together by drawing down a thin coating of a solvent based urethane adhesive over the skin of one laminate, sealing the other laminate skin to skin, and then placing in the oven at 110°C for 10 minutes to dry.

When one-inch wide strips were cut from the length of the laminate described above, it was found that the peel strips were variable in strength across the width of the construction. For example, a single 10-inch wide laminate gave peel strengths varying from 2.5 to 10 lbs/inch. The outside strips generally gave lower values than the inside. Further, the cloth frequently unravelled as the strips were pulled on the Instron tester. This resulted in a measurement of cloth tearing rather than adhesive peel. To remedy this problem and prepare a construction which yielded reproducible peel strips, a template was constructed (Figure 1). This template was placed on

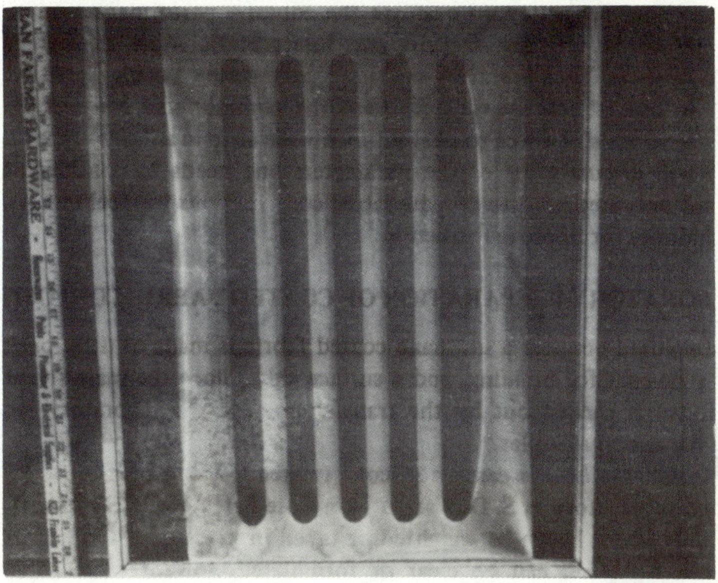

Figure 1. Template for Making Peel Strips.

the skincoat and the tie coat was drawn down over it similarly to the method described without the template (Figure 2). The template was lifted, leaving alternating one-inch strips having adhesive and no adhesive, after which the cloth backing was rolled into place and the laminate completed as above. To complete the preparation of the peel strips, two laminates were glued together as described above, after which the fabric could be cut through the center of the unbounded strip. Thus the test pieces have just one-inch width of bonding and about 0.5 inch unbonded cloth on either side of the bonded strip.

Figure 2. Hand Draw Down Using Template.

The construction of the template described above and shown in Figure 1 was from a sheet of Mylar film. We have used thicknesses from 2–5 mils. One-inch wide rectangular strips one-inch apart were cut out as shown and the Mylar was fitted into an aluminum screen frame.

The testing of the peel strips was performed in an Instron tester as follows. The ends of the strips had been deliberately left unglued and these were clamped into the jaws of the Instron tester at 2-inch separation, after which they were peeled at a crosshead speed of 20-inches/minute (Figure 3). The average values for peel strengths were more consistent with this method. For the laminate having the same composition as described above, the value obtained by averaging the forces where tearing occurred was 10–12 lbs/inch. Also the variability of the force required along the length of the strip, as shown by the Instron chart, was much less with these strips. The peel patterns obtained were all of the continuous failure type (4b). Also, no tearing of the cloth other than induced by the bond with the adhesive was evident as had frequently occurred with strips cut from the plain laminated construction.

Values of peel strengths obtained are not highly precise, and it is advisable to peel a number of the strips from a construction and average them. If only one strip from a hand made construction were pulled, one might report a value ranging from 5 lbs/inch to 12 lbs/inch. We are in the process of collecting statistical data in our

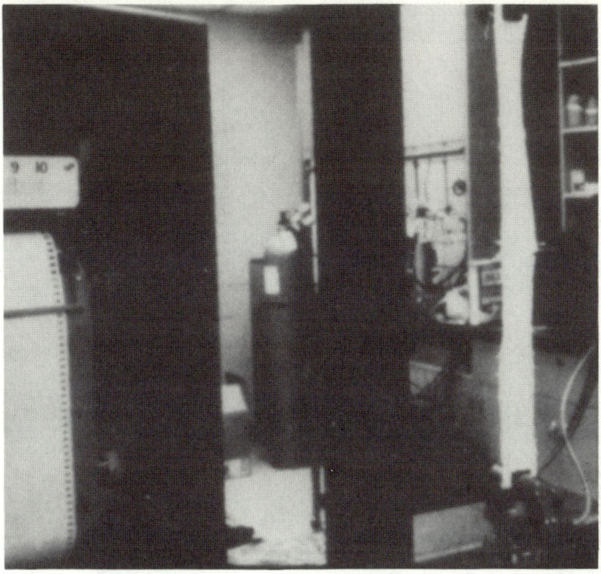

Figure 3. Peel Strip in Instron Tester.

own laboratory and co-operatively with other laboratories in order to determine a reasonable deviation for these results and thus the minimum number of strips which should be pulled per test.

Further, the peel strips thus obtained are useful for washing and dry cleaning tests, for humid aging or other aging experiments for which peel strength values are informative. Any unravelling of cloth edges does not affect results.

DISCUSSION OF VARIATIONS IN TESTING

Some of the factors we believe should be noted during the testing procedure are presented. The drying procedure for the preparation of the skin coat on release paper is important. It should provide a coating with essentially no residual volatiles since these can change the adhesive bond. Further, when drying the tie coat, not only is complete drying important, but also the rate of drying can affect results. For an aqueous tie coat, one must expect almost all of the water vapor to pass through the cloth, and thus drying rate can be controlled by temperature, humidity, weave of cloth and formulation. Complete gluing of the two skin coats is needed in order to prevent inadvertent failures in peeling. It may be necessary to consider the alternate of controlled heat sealing where an effective glue cannot be found. Even with all precautions, there is more variability in hand preparations than under equilibrated machine conditions.

Table 1. Peel Strength Variations with Skin Coats
Conditions: Cloth-1.21 heavy napped sateen

Hand Cast Skin Description	Latex Tie Coat av. lb./in. 2 mil template Latex A	2 mil template Latex B	Non template Latex A
Unformulated Lacquer (nondiscoloring Witcobond®)	7	4	6-7
Same as above + 2% silicone dispersant and/or pigment	5	4	3
Unformulated Estane® (discoloring type)	8	-	-
Same as above + 2% silicone dispersant	2	-	-
Calandered thermoplastic (nondiscoloring Tuftane®)	0	0	-

After evaluation of a great many samples, relative peel strengths were meaningful when conditions were controlled. Table 1 shows some effects of variants in urethane skin coats. Experimental data is presented on tie coat performance of two different urethane latices. Latex A is a 50% solids dispersion in water, free of solvent, containing thermoplastic particles ranging in size, 0.5–2 microns, based on a discoloring isocyanate. Latex B is a 33% solids dispersion in water, free of solvent, containing particles ranging in size from 0.1–1 micron, based on a non-discoloring isocyanate. Both of these latices are low viscosity and require thickening to about 15,000–40,000 cps. for good coating flow properties. Peel strength constructions were prepared using a 1.21 heavy napped cotton sateen and the skin coats as shown. Peel strengths are affected by processing aids on the surface of the skins and the nature of the composition. Some data was available which showed that peel strengths taken with and without the template were generally in agreement, but the values without the template were more variable.

In Table 2, peel strengths are compared for the same skin, but showing variants in the cloth. Better constructions can be made with heavy napped sateen than a thinner lighter napped cotton. Also, for this napped fabric the adhesion is about 15% better in the warp direction than the fill direction. Further, we have shown that the thickness of adhesive application is closer to optimum when using the 5 mil template than when using a thinner or thicker application. A more detailed and theoretical discussion of fiber adhesion has been presented by Wake [8]. Failures in the skin phase can occur, but normally available polyurethane materials are sufficiently stronger than the adhesive bonds when properly prepared and glued to-

Table 2. Peel Strength Variations with Cloth and Latex Thickness

Conditions: Skin-Unformulated Lacquer (nondiscoloring Witcobond®)

Cloth Description	Wet Thickness (mil)	Peel (av. lbs./in.)	
		Latex A	Latex B
1.21 heavy napped sateen	10	7	4
warp direction	5	9	6
fill direction	5	8	5
	2	7	4
510 low loft cotton	5	3	3

gether. The expected failure in a coated fabric composition should be between the tie-coat and fabric, usually tearing the napping. Failure between the tie and skin coats should be reported as a different type of result and in the systems we evaluated, seldom gave more than 1–2 lbs peel strength. Since we have been unable to control completely this alternate failure mode with some combinations of components, we believe that this is not so much related to the polyurethane polymers in tie and skin coats, but to the formulation and applications conditions when preparing such constructions.

Additional variation factors in the formulation and handling of the latex itself are shown in Table 3. Interfacial drying rate and several other physical effects are interrelated in the practical bonding of coated fabrics and thus are difficult to separate and demonstrate. Aqueous tie-coats at about 50% solids dry surprisingly rapidly, and in general adhesive performance will improve with somewhat slower drying. Even without well-defined experimental controls, we have seen improved peel strengths with the same polyurethane tie-coat at lower solids content, finer particle size range, using dampened cloth, and with better viscosity control in the tie-coat. Further, air-frothing of aqueous tie-coat adhesives introduces additional parameters to change mechanical bonding behavior, but these have not been covered here.

The improved test procedure proposed here should offer the means of developing improved applications conditions for preparations of coated fabrics. It is hoped that by future cooperative efforts in working out testing methods, better reproducibility and correlation will be achieved.

REFERENCES

1. J. S. Maddams, *J. Coated Fabrics 3*, 164 (1974).
2. British Standards Institution, London, Specifications for Coated Fabrics for Upholstered Furniture, BS 2601, 2602, BS 4216: 1970, Specifications for Coated Fabrics for Coated Fabrics for Protective Clothing BS 3546: 1970.

Table 3. Peel Strength Variations with Latex Variants
Conditions: Cloth-1.21 heavy napped sateen
Latex tie using 2 mil template

	Description of Variant	Peel (av.lbs/in) Latex A	Latex B
1)	Thickener-hydroxypropyl methylcellulose	7	4
2)	Same thickener plus 3 weeks or more aging	8	-
3)	Same thickener + adhesion promoter	7	-
4)	Same thickener, adhesion promoter + silicone skin	7.5	-
5)	Same thickener + 1% toluene	4	-
6)	(1) wet after washing test	0.5	4
7)	(1) wet from perchloroethylene test	*	*
8)	(1) redried one cycle	7	4
9)	(1) redried after 5 cycles	5	3
10)	Polyvinylpyrollidone type thickener	7	8.5

*dependent on skin composition

3. Chemical Coated Fabrics and Film Association, Chemical Coated Fabrics and Film Standard Test Methods, 5th ed., New York, 1972.
4. R. L. Patrick, ed., Adhesion and Adhesives, Marcel Dekker, Inc., New York.
   a) J. R. Huntsberger, Vol. 1, Chap. 4, 119 (1967).
   b) J. L. Gordon, Vol. 1, Chap. 8, 269 (1967).
   c) H. T. Patterson, Vol. 2, Chap. 6, 261 (1969).
5. G. C. Kantner, Adhesives Age *10* (6), 22 (1967).
6. D. Popplewell and L. G. Hole, *J. Coated Fabrics 3*, 55 (1971).
7. J. R. Russell and D. W. Ward, AATCC Symposium on Coated Fabrics Technology, March, 1973, p. 30.
8. W. C. Wake, *J. Coated Fabrics 3*, 84 (1973).

A. D. FUSSA AND S. V. NABLO

Energy Sciences Inc.
Burlington Road
Bedford, Mass. 01730

THE ELECTRON CURING OF (URETHANE) FABRIC COATINGS*

(Received September 19, 1974)

ABSTRACT: This article describes electron curing of urethane compounds. The method allows the use of high solids, single component urethanes. Line speeds of 250 feet/minute may be realized with this new technique. Process innovations such as curing the coating through the release film or using conventional transfer paper for casting a textured coating, are possible.

KEY WORDS: Electron resin cure, urethane fabrics, free radical cure.

INTRODUCTION

A NUMBER OF comprehensive review articles on electron curing of surface coatings, and the interrelation of the paint technology, chemistry and physics of the process are available in the literature [1–3]. This paper will not attempt to survey the process, but will outline recent developments in the energy sources used for industrial electron curing and some initial results for urethanes.

It is worthwhile to review the basic features of this energy source and curing process, compared with those also under consideration as replacements for thermal curing. The most important of these are:

(1) The energy emerging from the processor is *highly directed* – there is little inefficiency involved in the need for redirection or focusing of the energy. The energy is in the form of high speed electrons, and the optics of a well designed generator assure that most of the energetic particles reach the product. This represents a very significant advantage over the "undirected" tube, lamp or oven.

(2) Electron curing results from free-radical induced polymerization of the urethane polymer system; the same coatings used for ultraviolet curing [4] are useful for this curing technique. The significant difference in the processes is that electron curing requires *no photo-initiators* or *photo-sensitizers* to enhance

*This paper was originally presented at the American Chemical Society Meeting, Coated Fabrics Symposium, Atlantic City, NJ, September 11, 1974.

free-radical generation, so that stable, lower cost coating systems can be used. The reason lies in the energy levels of the sources: ultraviolet light (photons) deliver ~ 3 electron volts upon absorption, the electrons arrive at the coating with ~ 10^5 electron volts of energy.

(3) *The source generation efficiency is high* because it is not a "radiative" source (compared with gas discharge or thermal systems). Roughly 70% of the electrical power used in the electron generator is available as directed process energy for ambient use. This is at least a factor of 10 better than the available (radiation/thermal) alternatives.

(4) *The product coupling* of the electron energy *is efficient and controllable.* There is little substrate wastage – the curing station can be operated in a slaved mode with instant start-up or shut-down . . . quite a different situation from conventional radiative (infrared or uv) systems.

Many other comparisons could be cited but these four are the most significant; in fact they make a number of new curing processes possible because each involves, to some degree, a new process control parameter. We shall now examine the ability of the curtain processor to bring these advantages to industry in a practical way.

THE ENERGY SOURCE

Several views of a curtain processor are shown in Figures 1 to 3. As shown in the schematic of Figure 1, it consists of two concentric cylindrical tubes, the outer one of which serves as the processor housing, while the inner one defines the electron source which is held at a high negative voltage. The inner assembly is excited electrically from an integral power supply as shown in Figure 2, or by means of a cable-connector assembly as shown in Figure 3 [5]. In either case, the processor head is maintained at a modest vacuum by means of a small mechanical pump (Figure 2) or a purely electrical pump (Figure 3). There are *no* exposed voltages in the system so that the operator hazards or electrical interference (rfi) often associated with high voltage machinery, are completely eliminated.

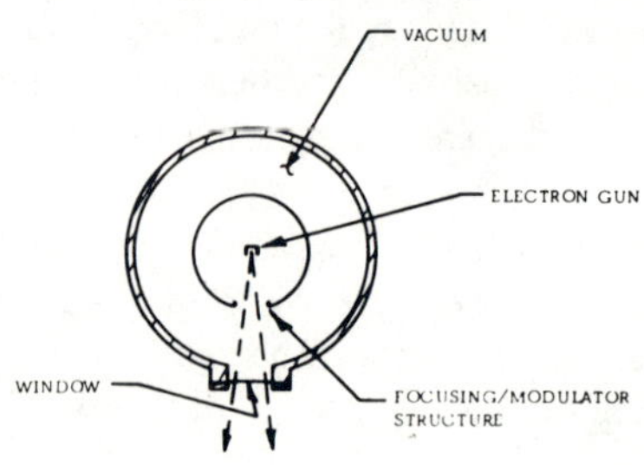

Figure 1. Curtain processor schematic.

Figure 2. The CB 150/125/50 Electrocurtain^{TM} (48 inch) processor with modular product conveyor.

Figure 3. Cable fed Electrocurtain^{TM} (CBS 175).

The processor head or tube works electrically as a grid controlled triode. The head itself is typically 8–10 inches in diameter, and contains a metal foil window (typically .001 inch (25 μ) of Al alloy) along a portion of its length to permit the flow of the energetic electrons from the tube to the product. This window effectively defines the area continuously illuminated by the energetic curtain over the product line . . . normally a section ~ 2 inches and of full web/product width.

For the Electrocurtain™ units shown in Figures 2 and 3, operation is typically at 175 kilovolts which will provide a useful penetration of 15 mg/cm^2 or 6 mils (150 μ) for a unit density coating. This has proven to satisfy a wide spectrum of industrial (protective, decorative, adhesive) coating curing applications.

It is obvious that some energy loss occurs in transmission of the curing flux through the window [2, 5]. For the units shown, this turns out to be about 15% at 175 kilovolts and this power is dissipated through conductive cooling of the window. This loss is the result of "slowing down" of the electrons in the foil; only about 5% of the electrons themselves actually stop in the window at this energy. As a consequence, the processor works at quite high electrical conversion efficiency, considerably in excess of 60% depending upon the size of the unit. Since over 80% of this emergent energy can be delivered to the coating for a well designed system – overall power efficiencies of above 50% characterize the curing process. This is a critical factor in preserving the low temperature properties of the cure; i.e. that the intrinsic advantages of radiation curing are not lost by the presence of high levels of thermal (infrared, convective) heating due to the low electrical efficiency of the energy source. This has frequently turned out to be the case with various forms of ultraviolet apparatus, for example.

PROCESSING CONSIDERATIONS

The Energy Investment for Curing

For the free-radical induced urethane systems available for use today, the energy required to accomplish a full cure is around 10 calories per gram of coating. Since the specific heats of most of these coatings are near 0.5 (cal/gm/°C), the maximum temperature elevation accompanying the process is ~20°C. This is, of course, one of the prime advantages of the technique . . . a room temperature process with lower volatiles so that, for the first time, a fully self-contained curing system is practicable.

Curing Speed

Since the curing process is clearly power limited, the processor output is normally tailored to the web speed required; i.e. to the coating throughput presented to the curing station. For a processor of the type shown in Figure 2, the curing station can provide power levels of 200 watts per inch (80 w/cm) of curtain length (or about 50 cal/second/inch of length), with which coatings of the type described above can be cured at line speeds of 250 feet/minute. Decorative coatings (inks) are available which cure at 1/10 that energy investment, or at 1 cal/gm, so that curing at line speeds of several thousand feet per minute is practical with the existing machinery. Typical speed:power curves for typical coating cure requirements are shown in Figure 4. Further extension of process speed is practical with elongation of the processor window along the direction of web motion, or by the use of tandem units.

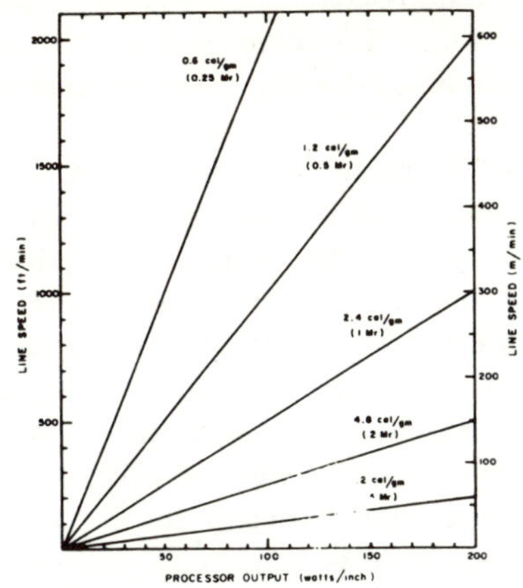

Figure 4. Curtain processor performance as a function of cure requirements.

Coating Penetration

Some calculated coating curing data are presented in Figure 5 for the performance of a 175 kilovolt processor with coating thicknesses of 25, 75 and 125 μ. The curves show the uniform manner in which the curing energy is deposited in the coating (in this case a pigmented acrylic on steel), regardless of coating thickness for the range shown. The curves demonstrate a treatment uniformity in depth of better than ± 10%; for example, the 150 keV output provided by a unit such as that shown in Figure 3 will cure a coating of up to 8 mils (200 μ) thickness. If thinner coatings are used with the same level of operation, an increasing percentage of the energy will be delivered to the substrate, as the coating thickness decreases (see Figure 5). Nevertheless, the processor will cure coatings from 0–200 μ at the same speed . . . the coating thickness variation will simply reduce the power efficiency of the process as the processor output is clearly most efficiently coupled to the coating for a given operating voltage. Since the overall efficiency of the curing process is already at least a factor of 100 that of its thermal counterpart, or at least several times that of the ultraviolet route [6], these variations are of secondary importance.

In summary, the curing station is operated to provide a curing capability with a good margin of safety for variations in either the application or substrate uniformities. Precise control of curing penetration, although available, is unnecessary except where it is desirable to control the degree of cure at the substrate interface for reasons of adhesion, film release, etc. This control parameter (penetration) offers new latitude to curing, and to new product development.

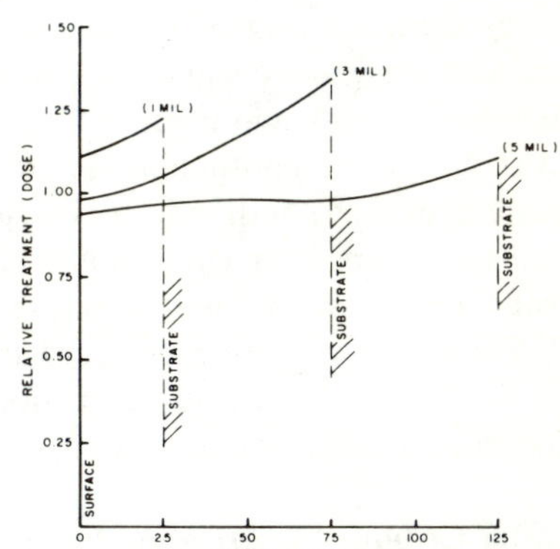

Figure 5. Treatment uniformity on a steel substrate for a 175 kilovolt processor and an acrylic – 5% TiO_2 coating.

Product Handling and Safety

There are a number of safety advantages in the curing of coatings by this process: notably the elimination of high solvent concentrations with the elimination of the toxicological, fire and explosion hazards associated with the conventional technology. The energy source itself does generate ionizing radiation which must be contained (shielded) so as to eliminate any hazard in the operating environment.

The units discussed earlier are designed to operate in an "unrestricted area", defined by OSHA [7] to include "any area, access to which is not controlled by the employer for purposes of protection of individuals . . . ". How this is done is shown schematically in Figure 6, in which the direct coating and curing of urethanes on a flexible web substrate is illustrated. The web is taken across rollers, after passing into the shielded web handling assembly. In this manner, the entrance and exit apertures leading to the process zone do not permit line of sight access to the critical, radiation filled region under the electron permeable window. For the processors described earlier, these web handling assemblies typically involve a total shielded distance of under 3 feet along the direction of product motion. They are fully interlocked to provide full access to the process zone, for cleaning of the processor assembly, window service and web threading.

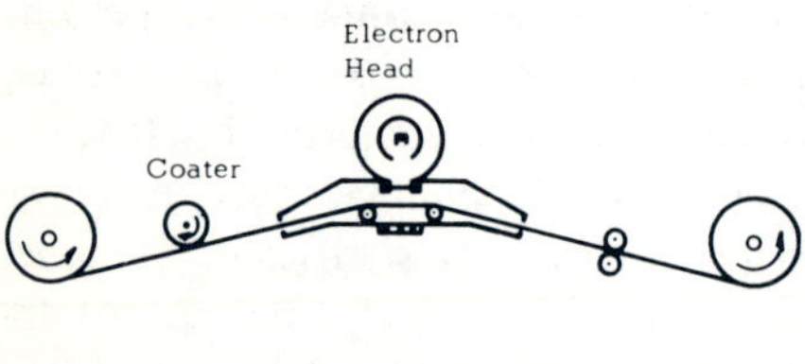

Figure 6. Direct coating urethane line.

The shielding assembly shown in Figure 6 serves two roles: it defines the treatment region and accommodates continuous product handling while confining the radiation flux generated in the curing zone. In addition, it serves as the boundary between the controlled inert atmosphere, frequently required for "tack-free" coating curing, and the ambient environment. This second requirement for high speed free-radical induced curing typically involves the reduction of oxygen concentrations below 1000 ppm in the curing zone, so that slender channels of low gas conductance are advantageous for the web application, as shown in Figure 6.

Although these continuous electron processors are fully interlocked to provide fail-safe operation, both industrial scale and laboratory type machinery are provided with back-up environmental monitors. In this manner, a completely independent picket monitor is available to ensure that the installation is operating well below OSHA limits. It will preclude processor operation if any portion of the system has been altered so that excessive environmental radiation levels are generated in the immediate environment of the unit.

Transfer Casting of Coatings on Fabric with Electron Curing

The flexibility and speed of the curing station permits process innovations impracticable with more conventional (e.g. moisture cure) urethane coatings. Two of these will be outlined to provide an indication of the impact these can have on the technology of fabric coating.

The first technique involves the curing of the coating *through* the release film, in this case mylar. The urethane coatings were first laid down by knife or drawbar on a one mil mylar film, then "set" with a single pass under the curtain processor. A tie coat is then applied to the partially cured film, the fabric tied or padded into it with a roller, and the entire composite then cured by a single pass treatment through the mylar release film.

The second technique is shown schematically in Figure 7 and utilizes conventional transfer paper for casting a textured or smooth coating. In this case, the transfer paper is coated with the urethane and partially cured. As in the first case, the tie coat is then laid down and the fabric padded in, but the composite is now cured through the backing or supporting fabric. This transfer casting technique permits the use of heavy release paper (e.g. 10 mils thick) which would tend to absorb most of the electron energy if curing were conducted from the release paper side. Using the approach of Figure 7, relatively thick coatings can be cast on a release paper of arbitrary ("electron absorptive") properties – for example metallic foils could be used . . . or the cast films could be released from metal casting rollers. The tie coatings are designed to penetrate the backing fabric to a predetermined thickness e.g. 0.5–4 mils, depending upon the end use of the product. In this way, the relatively "open" fabric – insofar as the penetrating electron flux is concerned – can then be used to transmit the curing energy to the adhesive or tie coat.

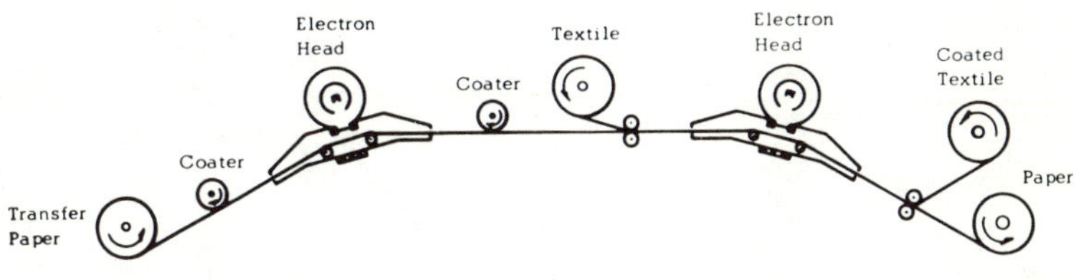

Figure 7. Urethane transfer casting line.

This technique has been used with fabrics of from 6–15 ozs per square yard, (20–50 mg/cm^2) using electron energies in the 120–160 keV range. Skin or cast film thicknesses of 0.5–5.0 mils are typical with the adhesive or tie coat thicknesses in the range mentioned.

RADIATION CURED URETHANE COATING PROPERTIES

The resins used in formulating these coatings are single component blocked aliphatic-acrylic modified urethanes. In some instances, aromatic urethanes have also been used, particularly for the tie coat where reduced durability of the tie coating is acceptable and viscosity control essential.

Some typical electron cured coating properties are shown in Table 1, as determined using ASTM D2370-68 on stripped films. The tensile strength at break is shown in Figure 8 as a function of delivered dose (1 megarad = 2.4 cal/gm of coating) while the modulus of elasticity at 100% elongation is shown in Figure 9 for these films, both clear and pigmented. In the case of the pigmented films, a 30% pigment loading was used. All films were 1.5–2.0 mils dry thickness and were cured at 100 feet per minute.

Table 1. Electron Cured Urethane Film Properties

Dose (Mrads)	Tensile at Break (psi)	Modulus at 100% (psi)	Tear Strength Pounds/Inch
1	890	300	35
2	1400	570	40
4	1760	1350	130
6	1860	1360	110

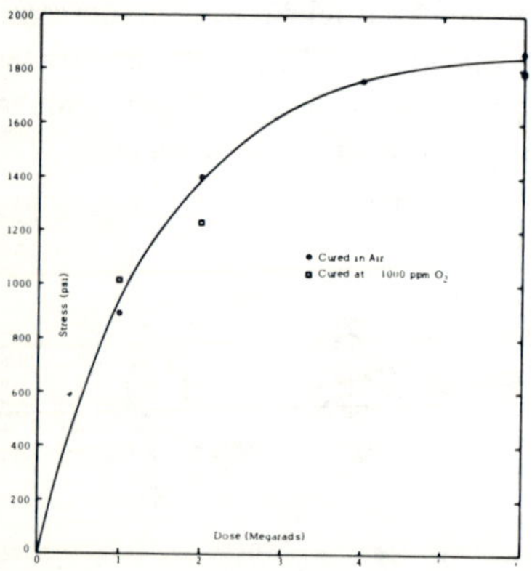

Figure 8. Urethane tensile strength at break. (ASTM D2370-68)

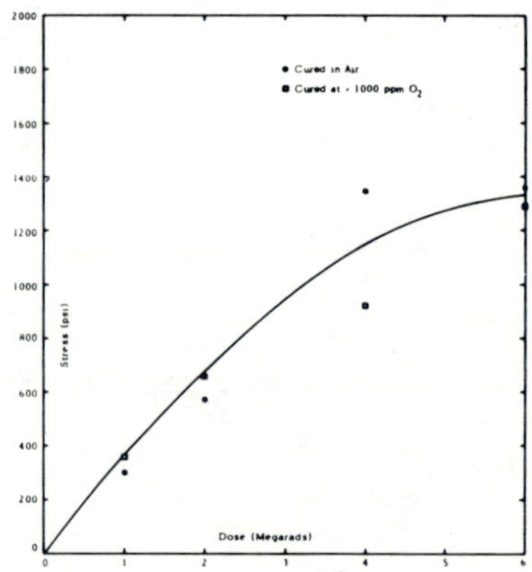

Figure 9. Urethane modulus of elasticity at 100% elongation. (ASTM D2370-68)

For the particular resin used in the data of Table 1, blocking occurred at the 1 megarad treatment level, with improvements in solvent resistance and dry cleanability with treatment level up to 4 megarads.

Some comparative data for ultraviolet cured *clear* films are shown in Table 2 using 1.5–2 mil films of the same resin. In case 1, a 2 lamp system at a 200 watts/inch power level was used with a single 0.15 second exposure (an equivalent line speed of 50 feet/minute). Case 2 utilized a second pass or 2 × 0.15 second exposures in the same system. Case 3 utilized a 3 stage Linde system at a line speed of 50 feet per minute.

No data are shown for the pigmented films in Table 2 as these could not be cured with the ultraviolet technique due to the great opacity of other than clear coatings, to radiation in this wavelength region.

Table 2. Comparative Data For Ultraviolet Cured Films

System	Tensile at Break (psi)	Modulus at 100% (psi)	Tear Strength Pounds/Inch
1	1120	610	50
2	2420	970	95
3	1540	600	130

Comparative studies of thermally cured films vs radiation cured films have also been conducted in this work. Comparable tear strength, Balley flexure resistance, Taber abrasion and Wyzenbeck break rubbing resistance have been demonstrated – with improved blocking, dry cleanability and modulus of elasticity (at 100% elongation) offered by the electron cured films.

CONCLUSIONS

Electron curing of these urethane compounds permits the use of high solids, single component systems in the coated fabric industry. The process is suitable for clear or pigmented systems with minor modification of the handling equipment already used for the conventional process. The low heat buildup in the coating permits the use of thick films and relatively labile substrates or backing films, such as unsupported vinyl, paper, etc. Most importantly, the curing station is no longer the pacing element in the line, nor does it tie up a large product inventory. Since the curing is "all electrical", precise monitoring of the process, and its control, are possible.

The large energy advantage demonstrated by the process, coupled with its reduced dependence on (if not elimination of) solvents, will ensure increased use of electron curing in the coated textile industry. The scope of its application will now be paced by the available free-radical cured coating systems suitable for industrial use.

ACKNOWLEDGMENTS

The authors are indebted to the Hughson Chemical Company of Erie, Pennsylvania for their cooperation in this program. In particular the work of Norm Adams and Jim Fonda in providing the resin systems, and in comparative film characterization, were essential to the success of this study.

REFERENCES

1. W. H. T. Davison, "Process Characteristics of Electron Beam Curing," Journal of Oil and Colour Chemists Assoc., *52*, 946, 1969.
2. D. R. Bailey, and A. Wright, "Electron Beam Curing," Paint Technology, 12 Sept. 1971; 23 Oct. 1971; 11 Nov. 1971; 23 Dec. 1971.

3. S. V. Nablo, and A. S. Denholm, "The Status of Electron Beam Curing (1971)", Proc. Nat. Coil Coaters Assoc., 1971 Annual Mtg.; NCCA, 1900 Arch St., Philadelphia, PA 19103.
4. S. H. Schroeter, "The Ultraviolet Curing of Coatings," *Non-Polluting Coatings and Coating Processes*, ed. J. L. Gardon and J. W. Prane, Plenum Press, NY, 1973, pp. 109–134.
5. S. V. Nablo, J. R. Uglum, and B. S. Quintal, "Electron Beam Processor Technology," ibid, 179–194; the Electrocurain^{TM} units are covered under U.S. patents 3,702,412 and 3,745,396; other patents pending.
6. J. G. Calvert, and J. N. Pitts, Photochemistry, Chapter 7, Experimental Methods in Photochemistry, John Wiley & Sons Inc., New York, 1966.
7. Occupational Safety and Health Standard, Federal Register *37* #202, Part II, 22158, Oct. 1972.

IRVING TASHLICK

Alva-Tech Inc.
Bradley Beach, NJ

SOME CURRENT INDUSTRIAL PRACTICES IN THE MANUFACTURE OF URETHANE COATED FABRICS

(Received October 5, 1974)

KEY WORDS: Urethane fabrics, tie coats, hydrolytic stability.

AROMATIC vs ALIPHATIC URETHANES

THE COST OF aliphatics are higher than aromatic urethanes and they also tend to have slightly less solvent resistançe. European coated fabric manufacturers tend to use more of the aromatic than do the U.S. manufacturers. Apparently their standards in regard to light stability are not as high. Even in the U.S., fabrics for upholstery and shoes are often based on the aromatic isocyanates.

TIE COAT RESINS

Originally the adhesive or tie coat urethanes were similar in composition. Now resins are engineered specifically for use as adhesives or tie coats. They tend to be softer resins and cheaper. There are also lower molecular weight types available in higher solids concentration. There is some controversy as to whether an aromatic isocyanate based adhesive used with an aliphatic isocyanate base urethane film causes the resulting coated fabric to have poor light stability. It has been shown with some systems that light stability is indeed poor, but there are recurrent claims that good results can be achieved.

In some countries acrylic latices are used as the adhesive, but in the U.S. this practice is not widespread, although good results are claimed.

Urethane latices have also been promoted for use as adhesives, particularly in frothed versions.

ADDITIVES

Pigments

Most pigments are used as preground high solids pastes. Polyesters or plasticizers are used as grinding or dispersing media. Some pigments can be ground in the solvents, or the lacquers can be used to make concentrates. Preground coated pigments and dyes are gaining popularity as a means of avoiding the grinding step.

Silicones

Silicones have been important to the success of urethane coated fabrics. With many urethane resins it was not possible to obtain a good feeling, slippery, tack free surface particularly in the case of a glossy patent surface. The problem was often solved by using various silicones as internal additives or as topcoats applied to the finished coated fabric.

Resins

Certain vinyl chloride-vinyl acetate copolymers are compatible with urethanes. They have been used as additives to lower cost and increase hydrolytic stability. They also usually increase stiffness and lower physical properties. Other resins such as phenoxies, ethyl cellulose, acrylics and nitrocellulose are compatible, but are only occasionally used in small amounts to impart slip or adhesion.

Miscellaneous

Antioxidants are used to prevent yellowing at high temperatures and U.V. absorbers are used in aliphatic systems to improve light stability. These materials do not drastically improve the poor light stability of aromatic based urethanes.

Carbodiimides are claimed to improve hydrolytic stability, but it is doubtful that they are widely used.

TESTING OF COATED FABRICS

Table 1 lists the characteristics of widest interest in the many uses of vinyl and urethane fabrics. There are many tests which are highly specialized for each application. Many of these depend on the substrate and how it is sewn, cut or handled. In Table 1 only tensile strength is mainly dependent on the substrate, the others are mainly tests of the performance of the coating.

In vinyl coated fabrics many problems arise due to the plasticizers. Loss of plasticizer by extraction with solvent or water or through volatility lead to product failure by embrittlement and cracking. This is probably the most common cause of failure of vinyl coated fabrics. However, there are several tests which correlate reasonably well with performance. The best compromise of price and performance can be worked out and proper recommendations made.

In the case of urethane coated fabrics, there have been two areas of particular concern. One is the quality control problem of consistently achieving the same degree of adhesion to the pile. The standard test for adhesion does correlate well with performance and is used in many modifications as a control test.

Table 1. Tests Commonly Used for Vinyl and Urethane Coated Fabrics

1. Tensile Strength	6. Abrasion Resistance
2. Blocking	7. Cold Crack Resistance
3. Adhesion to Substrate	8. Flexing
4. Dry Cleanability	9. Water Vapor Permeability
5. Accelerated Light Aging	10. Hydrolytic Stability

The second area of concern is the prediction of hydrolytic stability of urethane fabrics. Table 2 shows the many kinds of test conditions which have appeared in the literature. The full immersion at 70°C appears to be the most widely used test condition. A standard test for hydrolytic stability of vinyls has been to immerse the fabric at 70°C for 14 days and then to run 15,000 cycles on a Newark Flex Tester, and then examine for cracks, peeling and delamination. This is a severe test for many urethane coated fabrics.

Table 2. Conditions Used for Hydrolitic Stability Tests

1. 100°C immersion in 0.12% UREA Solution
2. 120°C immersion in 0.12% UREA Solution
3. 15 PSI Steam at 125°C
4. 95% RH at 70°C
5. 95% RH at 97°C
6. 95% RH at 80°C for 12 days
7. 70°C immersion in water

There is no accepted standard test yet, and it is difficult to predict performance by the use of hydrolysis testing. In demanding areas such as shoes and upholstery, there have been serious quality problems. There is a need for two kinds of tests to solve this problem. A flex test after hydrolysis should be adequate for the user as quality control. There should also be a quantitative test such as loss of film strength on hydrolysis which the coater can use to check his top coats and adhesives.

IRVING TASHLICK

Alva-Tech Inc.
Bradley Beach, NJ

URETHANES IN COATED FABRICS: PROCESSES AND PRODUCTS*

(Received October 5, 1974)

KEY WORDS: Urethane fabrics, PVC fabrics, Coated fabric properties.

INTRODUCTION

AT THIS POINT in time, most of the yardage of urethane coated fabrics is produced by the transfer process. A solution of thermoplastic urethane containing about 30% solids plus pigment is knife coated onto a release surface, usually treated embossed paper. The paper is passed through an explosion-proof oven to evaporate the solvent to give a film 1 to 3 mils thick lying on the release paper. This film becomes the top surface on the coated fabric and provides the color, texture and wear characteristics. A solution of adhesive, usually urethane based, is applied to the surface of the film, and fabric, usually with a napped surface, is pressed into the tacky adhesive layer. The combination of fabric, adhesive, urethane film and paper is passed through another drying oven, cooled and the paper is stripped off. The resulting fabric is then often printed and topcoated.

The paper is saved and may be used several times before it must be discarded. The release surface can be smooth or embossed, and its exact texture and sheen will be mirrored by the urethane film.

This transfer process is especially suitable for systems containing large quantities of solvent, and where deep penetration of the adhesive into the cloth would result in undesirable stiffening and impart a poor hand to the finished product.

Urethane coated fabrics for tents, tarpaulins and rainwear are made by direct coating onto the fabric. Similar equipment to that described for the transfer process is used except that instead of coating the paper, more viscous higher solids solutions are applied directly to the fabric. There is more penetration down into the fabric and only tightly woven or treated textiles may be used for direct coating. Otherwise the resin solution will strike through the fabric rather than coat it.

*Presented originally at Urethanes in Coatings Seminar, sponsored by Program Division, Technomic Publishing Company, Inc., Westport, CT, October 2–3, 1974, Chicago, IL.

CURRENT STATUS OF PVC AND URETHANE COATED FABRICS

From an examination of Table 1, it can be seen that there has been a steady evolution of the improvement of properties of coated fabrics over a period of 200 years. During that time only five different basic polymers have each in turn captured the bulk of the coated fabrics market.

Table 1. Qualitative Ratings of Properties of Coating Resins Used to Produce Coated Fabrics

	Materials				
Property	Drying Oils	Rubber	Pyroxylin	Plasticized Polyvinyl Chloride	Poly-Urethane
Tensile Strength	Poor	Good	Good	Good	Excellent
Tear Strength	Poor	Fair	Good	Fair	Excellent
Flex Resistance	Poor	Good	Poor	Good	Excellent
Abrasion Resistance	Poor	Good	Excellent	Fair	Excellent
Elongation	Poor	Excellent	Poor	Good	Excellent
Light and Heat Stability	Poor	Poor	Poor	Good	Good
Hydrolytic Stability	Fair	Excellent	Good	Good	Fair
Solvent Resistance	Poor	Poor	Fair	Fair	Excellent

Oilskins were dominant for about 100 years before the advent of rubber coated fabrics. The rubber coated product held sway for over 70 years before pyroxylin coated goods took over. The heyday of pyroxylin lasted 30 years before being supplanted by PVC. PVC has been dominant now for about thirty years. Urethane coated fabric became commercially significant only five years ago, so one must conclude that viewed from an historical perspective it is still a young and undeveloped upstart.

From an examination of the kinds of PVC coated fabric and the kinds of urethane coated fabrics which are produced, a similar conclusion becomes obvious.

Vinyl coatings are available on a huge variety of substrates. Relatively cheap cotton knit fabrics are the most widely used substrate for such end uses as automotive upholstery, furniture upholstery, apparel, and luggage. Woven heavy-weight cottons such as twills and drill fabrics are used for automotive and truck upholstery and luggage applications, the more closely woven sateens are used in shoes, while cotton sheeting is used for wall covering. Non-woven fabrics are also employed in a variety of applications. Light weight non-wovens are used for shoe interliners, while the heavier weight non-wovens find applications in automotive uses, luggage and shoes. In short, a vinyl coating is put on any fabric which meets cost-performance requirements; and several types of substrates can compete in one kind of application.

A similar situation exists for the vinyl coatings themselves. They are widely produced in thicknesses from 5 to 15 mils in solid versions. Foamed vinyl coatings are available in the range of 6 mils to a full 1/8 inch thick.

When the large variety of substrates used are combined with numerous variations in coatings, it is apparent that vinyl coated fabrics represent an almost infinite variety of products which can be engineered and styled for particular applications. In addition, three different processes – calendering, extrusion, and casting – are used to make these products, with each one contributing some variation in product quality and appearance.

The situation with urethane coated fabrics is altogether different. In the industry today a urethane coated fabric indicates a very specialized product.

A highly napped fabric, preferably with a smooth dense pile is used, and it is coated with a 1 to 3 mil film which tends to sit on top of the pile. This construction gives urethane coated fabrics their characteristic soft, supple hand and a texture with a fine leatherlike break or crinkle. It also leads to some of the problems associated with PUCFs which will be discussed later.

Different weights and qualities of napped fabrics are used for different applications. Napped nylon knits have been used for gloves and women's stretch boots, napped sateens are used in shoes, while twills are preferred for upholstery and apparel. Except for some coated foam products which were used to make handbags and the direct coated fabrics used in rainwear, military and industrial applications, the overwhelming yardage of PUCF is a product which consists of a transfer coated thin film resting on a napped fabric. This specialized construction limits the kind of decorating and textures which may be used with PUCF. The reason this product was developed first with urethanes is that it could not be done previously since none of the other coating polymers had the unique combination of properties which was needed. This construction resulted in urethane coated fabrics initial success, but also resulted in some of the problems associated with them. Deeply embossed surfaces or extremely smooth surfaces are textures which cannot be obtained with the thin film over pile construction.

POTENTIAL URETHANE COATED FABRIC PRODUCTS AND PROCESSES

Table 2 lists the types of urethane systems which can be used to produce coated fabrics, while Table 3 lists the processes which can use these systems. Evidently urethanes have the potential to surpass PVC in the versatility of production methods which may be used. It would seem probable that eventually urethanes would compete with PVC in every type of use and process and follow the history of PVC development and perhaps even displace it completely.

Table 2. Available Polyurethane Systems

1. Urethane Lacquers
2. Reactive Solutions
3. Water Dispersions
4. Thermoplastic Pellets
5. Isocyanate Terminated Prepolymers
6. Two-Component Prepolymers

Table 3. Processes to Produce Urethane Coated Fabrics

1. Direct Coating
2. Transfer Coating
3. Calendering
4. Extrusion
5. Lamination

Two circumstances could speed this process. Over the years the gap in cost between vinyls and urethanes has tended to narrow as the urethanes become less exotic materials. It has also been reported that certain concentrations of vinyl chloride monomer may cause liver cancer. The ultimate impact of this finding is difficult to assess. It may simply make vinyls more expensive because of safety precautions which would become necessary to the production of monomer and polymer. It could make processing and fabrication more difficult and expensive if it is found that hazardous concentrations of monomers are generated and, improbable as it may seem now, even cause consumer rejection of vinyl products. At any rate, in the light of the preceding examination of the history of coated fabrics, it is difficult to visualize the enduring dominance of any single material.

If urethane coated fabrics are to increase in volume, products other than the thin film coated pile must emerge.

It is difficult to imagine that urethane coated fabrics will continue to consist mainly of thin films on napped substrates. The trend should be toward heavier films, utilization of the complete spectrum of substrates, and to the other processes available to coat them. For heavier films, the present solvent coating technology would be unsuitable.

Non-solvent processes are inherently more economical than solvent processes and they should become more important if urethane coated fabrics with heavy coatings evolve.

There are two kinds of non-solvent processes by which urethane coated fabrics can be manufactured. One type would be melt processes involving the use of thermoplastic resins; the other type would be the formation of urethanes from low molecular weight precursors. Calendering and extrusion represent the first type. As better thermoplastic urethanes become available these processes should become more important. Difficulties have been encountered in handling free urethane film in the lamination process, but satisfactory material has been made by these processes.

Another interesting route to urethane coated fabrics involves the use of two component non-solvent systems. Table 4 lists some of the advantages and disadvantages of a 100% solids non-solvent coating process. This process is analogous to the plastisol coating process used to make vinyl coated fabrics. In theory it represents the most economical route to manufacturing urethane coated fabrics. There have been several commercial and pilot plants operated using this process.

Table 4. 100% Solids Two Component Coating Process

Advantages:

1. Heavy Films in Single Pass
2. No Solvent Use
3. Analogous to Plastisol Process – May Use Knife Coaters and Ovens
4. Can Make Crosslinked Films
5. More Efficient Use of Energy and Materials

Disadvantages:

1. Cannot Make Films Below 3 mils
2. Requires Mixing and Metering Machinery
3. Systems Have Limited Pot Life

These systems consist of two prepolymers preferably liquid at room temperature. One prepolymer is isocyanate terminated and the other is hydroxyl terminated. The hydroxyl terminated prepolymer may be pigmented and catalyzed. The two prepolymers are continuously metered by pumps to a mixing dispenser and poured onto the moving web and spread coated. The mixing and metering machinery is similar to that used to manufacture urethane foam. The coating range may be either a plastisol or a urethane type and from this point on the process proceeds as in PVC or urethane transfer coating.

H. L. GEE

Managing Director
Elastomer Products, Ltd.
Leicester, England

THERMOPLASTIC URETHANES FOR EXTRUSION INJECTION AND BLOW MOULDING AND FOR TEXTILE COATING*

(Received December 27, 1974)

ABSTRACT: Thermoplastic urethanes behave as crosslinked rubbers when cold, but when heated to moulding temperatures, behave as linear non-crosslinked polymers which flow readily for moulding. The material is inert to most common solvents, has good abrasion resistance, and high elasticity. Being thermoplastic, scrap and sprues may be recycled, while in coatings, the material is finding a natural application in the construction of simulated leather products.

KEY WORDS: Thermoplastic urethane, urethane compounding, synthetic leather.

INTRODUCTION

THERMOPLASTIC URETHANES ARE one type of non-foamed urethane system isolated for consideration because of the fact that these members of the urethane family of systems are conventionally factory made and offered for sale after the fashion of other thermoplastic polymers. They are distinguished accordingly from other urethane systems typified by:

(a) 2 component urethane systems which react together to form plastic or elastic urethane solid polymer which cannot be melted without decomposition, and

(b) 1 component urethane systems containing a 'masked' reactive material which increases the molecular weight of the polymer by chain extension or crosslinking following unmasking during fabrication.

The thermoplastic urethanes secure for themselves a real, if at present small, market because they behave as crosslinked rubbers when cold, but when heated to moulding temperature behave as essentially linear non-crosslinked polymers which therefore flow readily for moulding purposes.

When applied as textile coatings, the crosslinking bonds, instead of being broken

*This paper originally presented at The Polyurethanes Symposium, University of Surrey, Guildford, Surrey, England, September 23–25, 1974, organized by The Polymers Paint & Colour Journal.

by thermal agitation, are broken by the use of an aprotic solvent/diluent system or formed as an emulsion with water to enable the fully crosslinked rubbery polymer to be reformed as a continuous film on the surface of textile materials.

MARKET

As indicated above, the otherwise unconnected markets of textile coating and thermoplastic moulding or extrusion can be served by the common product on the basis of thermo- or solvent labile crosslinks. When in service, the polymer is sufficiently crosslinked to virtually eliminate creep or cold set and yet has an even higher elasticity than a conventional vulcanised hydrocarbon rubber.

Coupled to this ease of fabricating a vulcanised rubber type of material is its inertness to most common solvents, an absence of embrittlement at low temperatures down to around $-50°C$, an outstanding abrasion resistance, and an outstanding high tensile strength. This combination of desirable properties offsets the admitted high price and ensures its use on a logical basis.

In the moulding field, there is the added convenience and economy of being able to re-cycle sprues and scrap whilst in the coating field the high tensile strength, being about twice that of natural rubber, coupled with a similar elongation at failure, find a natural application in the construction of simulated leather products, tear resistance proofed fabrics and poromeric synthetic leather.

Because of its speciality nature and relatively high price, thermoplastic urethanes presently account for only a small percentage of the total thermoplastic polymer market and a similar small share of the vulcanised rubber market. Nevertheless, the cost of production is still modest and some 70% of the total selling price is accounted by raw material costs. Clearly, therefore, price is closely related to volume production of these raw materials which in themselves start from low cost basic petroleum products. As most commodity plastics are already produced in very large plants, the economy of scale has already taken effect, whereas this is not yet the case with urethane thermoplastics. A narrowing of the price differential can, therefore, be seen as demand for increased performance provides more and more opportunities for sophisticated polymers typified by urethanes.

Here again, urethanes have an immediate advantage in that with some 3 basic ingredients, each of which can comprise half a dozen or more commercially economical products all combined in differing ratios, they can provide a wide and almost tailor made variation of properties without major influence on the base cost. At present, it is only the desire to have the majority of applications met by a standard product with well investigated properties which impels the manufacturer to limit his range. The chemist has an unequalled opportunity to exercise his imagination.

CHEMISTRY

The material is essentially a block co-polymer comprising a linear back bone made up of long flexible segments of molecular weight in the range 800–4000 having a polyester or polyether construction coupled in an ordered or random manner with a rigid low molecular weight alkyl aromatic urethane by means of further urethane links. This construction is invariably made by reacting an essentially linear saturated α,ω polyester diol or polyether glycol with a simple aliphatic or aromatic glycol and a diisocyanate.

The low temperature properties are very largely controlled by the long flexible segment as are the solvent resistant properties. The hardness and tensile strength depend mainly on the hard alkyl aromatic urethane segment.

A typical common thermoplastic urethane would be made by reacting ethylene glycol and adipic acid to form a polyester of molecular weight in the range 1000 to 5000 in the absence of an esterification catalyst and removing the water of esterification by continuous distillation. By control of temperature and avoding oxidation a glycol adipate polyester results having an average M.W. of about 2000 and having a good colour, low acid number, low unsaturation and a hydroxyl number and viscosity conforming closely to the average molecular weight.

A typical commercial product would be:

Elastophen	2000	(Elastogran GmbH)
Desmophen	2000	(Bayer AG)
Diorez	770	(Briggs & Townsend, Ltd.), etc.

$$nHOOC(CH_2)_4COOH + nHO(CH_2)_4OH \rightarrow [-O(CH_2)_4 \cdot O \cdot \overset{\displaystyle O}{\overset{\|}{C}}(CH_2)_4\overset{\displaystyle O}{\overset{\|}{C}}-]n + nH_2O$$

By variation of the glycol, e.g., 1,6 hexandiol, neopentyl glycol, butan 1,4 diol or using mixtures as well as by varying the diacid, e.g., phthalic, terephthalic, sebacic, etc., the melting point and viscosity can be widely varied.

Because of the high temperature of esterification and ester exchange, thermodynamic equilibrium is established and the molecular weight distribution follows a Flory probability rather than the narrower initial Poisson distribution. The time of the reaction is extended and may be up to 12 hours and there is a risk of metallic contamination. To avoid these problems and to produce a more regular repeat structure having better packing possibilities, a polyester can be generated by ring opening and chain extending a cyclic lactone as demonstrated by Carothers in his early work on polyesters. Only cyclic lactones with 4, 7 and 8 members will polymerise. Practically the monomers of interest are β-propriolactone derivatives, and ϵ-caprolactone. The former suffer the disadvantage that the polyesters derived from them depolymerise at a relatively low temperature and the parent compound is known to have physiologically undesirable properties. ϵ-Caprolactone is readily made by Bayer-Villager oxidation and rearrangement of cyclohexanone. The lactone can be polymerised to low M.W. material in the range 1000 to 5000 by use of a suitable polar starter and a base used together in a concentration related to the desired M.W. of the product polyester. It is normal to use a base and water or glycol as a starter and polymerisation proceeds along the following lines:

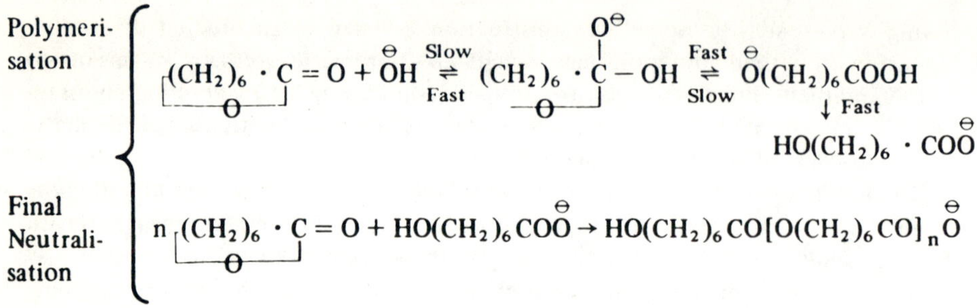

$$\mathrm{HO(CH_2)_6 \cdot CO[O(CH_2)_6CO]_n\overset{\ominus}{O} + \overset{\oplus}{H} \rightarrow H[O(CH_2)_6 \cdot CO]_{n+1}OH}$$

It is to be noted that there is no by-product water and the reaction time is shorter than for an esterification reaction. Whilst in theory the M.W. distribution should be of the narrow Poisson form, in practice due to ester interchange between different molecular weight species the final ratio Mw/Mn is nearer 2 corresponding again to the thermodynamic equilibrium distribution.

Typical commercial polycaprolactone diols having an average M.W. of 2000 are represented by:

Niax D560 (Union Carbide)
PCL 220 (Interox)

Alternative to these polyester diols polyether diols can be prepared by ring opening polymerisation of epoxides of which the 1, 2 and 1,4 types are typified by propylene oxide and tetrahydrofuran and which are widely used commercially. Ethylene oxide is occasionally used to 'tip' the resultant polyoxypropylenediol to increase the reactivity of the terminal hydroxyl group, although at the risk of reduced thermal stability of the ultimate polymer.

As with caprolactone, the polymerisation of propylene oxide is effected with a starter usually propylene glycol and potassium hydroxide, and the M.W. controlled by the amount of Propylene Oxide added.

$$\mathrm{HO \cdot CH(CH_3)CH_2OH + B \rightleftharpoons HO \cdot CH(CH_3) \cdot CH_2\overset{\ominus}{O} + \overset{\oplus}{HB}}$$

B = base

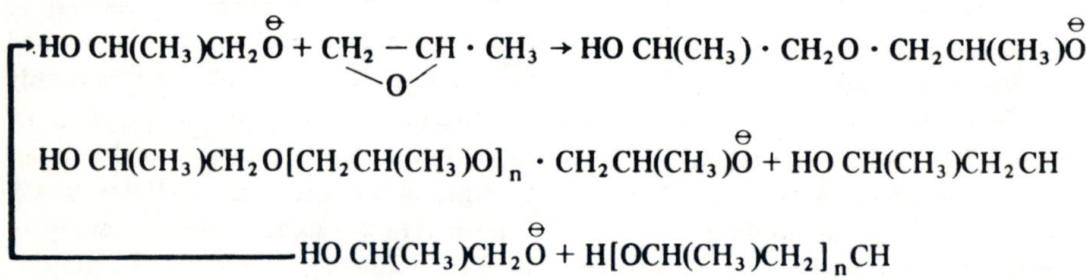

In parallel with this reaction there is the competing rearrangement of the propylene oxide to allyl alcohol and the chain extension of this to give a monoalcohol terminated with an unsaturated tail end. In addition chain termination by isopropylidene groups can also occur. These effects can result in low molecular weight urethane containing unsaturated groups. These side reactions increase as the diol M.W. increases and also result if the reaction is started at too low a temperature and is allowed to run away to too high a finishing temperature. Nitrogen base starters lead to further complications whereby the amine starter is quaternised at low temperatures and this species catalyses the reaction to give allyl alcohol.

The M.W. distribution follows the relatively narrow Poisson probability equation.

All base catalysed polyesters or polyethers must have the starting base neutralised or removed as if it were to remain it would uncontrollably catalyse the later reaction of the polyether diol with the diisocyanate and also cause the isocyanate to trimerise.

The classic route to urethanes involves the initial reaction of the 800–4000 M.W. polyether or polyester diol with a diisocyanate in substantial excess whereby only moderate chain extension occurs terminated by all terminal hydroxyl groups being capped by the excess diisocyanate. The resultant mixture is termed a pre-polymer. This pre-polymer being isocyanate terminated is then reacted with a stoichoimetric quantity of simple glycol or diamine to produce a high molecular weight urethane. It has subsequently been found that a product with substantially the same physical properties results from mixing all three ingredients together at once provided the polyether or polyester and the chain extender react at substantially the same rate. This is true for glycols, but not for diamines which react too fast and a polyurea precipitates out.

The choice of diisocyanates is very wide in so far as there is an enormous number of diamines capable of phosgenation. In commercial practice, the choice is much narrowed down and the major tonnage of urethane rubber is based on about 5, viz:

Aromatic: 2,4 and 2,6 toluene diisocyanate (TDI)
44′, 24′ and 22′ methane diphenyldiisocyanate (MDI)
1,5 naphthalene diisocyanate (NDI)

Aliphatic: 1,6 Hexamethylene diisocyanate
Hydrogenated MDI

(For speciality coatings, adhesives, etc., many more exotic aliphatic, alicyclic, alkyl-aromatic diisocyanates are used.)

The above aromatic diisocyanates are generally much more reactive and less volatile than the aliphatic ones and are, therefore, to be preferred unless the special characteristics imparted to the urethane by the flexibility and lack of electronic resonance of the aliphatic diisocyanates are required.

Diisocyanates are produced by phosgenation of the corresponding diamine. In the case of MDI this is the condensation product of aniline and formaldehyde which rearranges (Beckmann) under the influence of a strong acid:

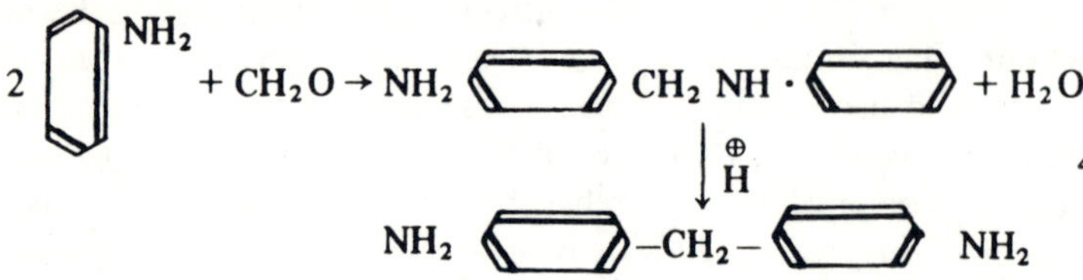

44′ MDI

Depending on the temperature, solvent system and pH varying amounts of the isomers,

42′MDA

and

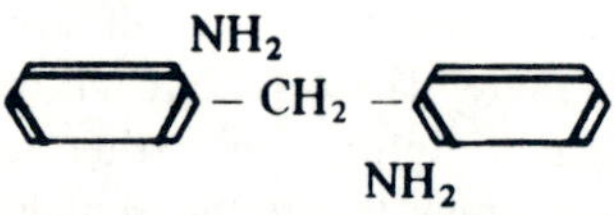

22′ MDA

are formed together with a variety of oligomers. According to the manufacturing process the latter are separated at this stage or subsequently after phosgenation of the crude MDA mixture to give a crude MDI mixture.

This latter method is commonly practised because of the large market for crude MDI for making rigid and semi-rigid urethane foam. High vacuum distillation in a thin film evaporator enables the stripping off of a small percentage of mixed MDI isomers (usually having an isomer ratio of about 92% 44′, 7% 2′4 and 0.8% 22′). Due to traces of residual phosgene and some chlorinated material, there is usually a small amount of readily hydrolysable chlorine in the distilled MDI which can later make the urethane sensitive to hydrolysis. This together with traces of iron, nickel, etc., which can have a variable catalytic effect make it desirable for these impurities to be closely controlled if used to make urethane elastomers. It is to be noted that only the 44′ isomer will pack well and therefore enhance the hardness of the rigid short segment of the poly-urethane. For the highest quality elastomers, therefore, the amount of 24′ and 22′ isomer must be minimised.

The 24′ isomer has for steric reasons a slight difference in the reactivities of the two isocyanate groups. Otherwise MDI reacts equally fast at either end and because of the isolation of the two resonant aromatic rings by the methylene bridge the reaction of the isocyanate group at one end of the molecule has no deactivating effect on the isocyanate group at the other end. This has the disadvantage that when making pre-polymers from MDI there is a greater risk of gelation. However, there is reduced risk of unreacted monomer remaining, and this taken in conjunction with the low vapour pressure of MDI makes MDI based pre-polymers attractive.

Being aromatic, the pooling of π electrons leads to enhanced reactivity of the isocyanate group. This is readily apparent when the ring is hydrogenated and the reactivity falls dramatically. This effect can only be offset by the inclusion of negative groups in the ring typified by $-NO_2$ and $-SO_2NH_2$, halogen, nitrile, etc. However, these substituents bring other problems and it is preferable to rely on more active catalysts.

By a similar process TDI is made by phosgenating mixed toluene diamines whose isomer distribution is determined by the nitrating conditions but usually results in about 80% 2,4 and 20% 2,6 with insignificant amounts of other isomers.

TDI is much more volatile than MDI and whilst easier to purify it presents more serious health hazards, particularly as isocyanates in general lead to individual sensitisation to their sternutatory and lachrymatory effects.

By fractional distillation to produce the pure 2,4 isomer a residue grade of TDI is offered having an isomer ratio of 65:35 which, being slightly less expensive, is used for the preparation of lower quality elastomers.

Because of steric hindrances, the ortho isocyanate group is considerably less reactive than the para group (referred to the methyl group) and further as both are coupled electronically through the aromatic π electrons, it follows that the para group reacts rapidly and deactivates the ortho group, which being also sterically shielded, then reacts comparatively slowly.

The net effect is that prepolymers prepared using excess TDI are very easy to make in a controlled fashion but nevertheless may contain up to 5% unreacted monomer which is difficult to remove and presents a significant health hazard. A further point which must not be overlooked is the temperature dependence of the relative reactivities of the two isocyanate groups – at higher temperatures the difference is much reduced and it is possible for the reaction to run away and gel, particularly in the presence of traces of alkaline catalyst, which by triggering off the competing trimerisation reaction, can substantially increase the reaction exotherm so as to fall outside the control of the cooling system.

Nevertheless, TDI pre-polymers are probably still the most widely used particularly for in-plant conversion to volume produced polyether foam.

The reduced reactivity of the pre-polymer terminal isocyanate groups makes it essential to catalyse the final urethane reaction and possibly to apply heating in addition to ensure a complete reaction and full cure.

Following the preparation of a pre-polymer, now of medium molecular weight terminated with isocyanate groups, these are coupled together using a simple bifunctional molecule. A small molecule is preferred because of the reduced mobility of the pre-polymer molecules and to preserve the hard rigidity of the isocyanate/small molecule segment which alternates with the flexible polyester or polyether segment. Further this small molecule needs to be reactive to chain up the pre-polymer as completely as possible without being in large excess. For this reason short chain glycols, diamines or bishydrazides are preferred. Bisphenols are not very reactive nor are the resultant urethane groups thermally stable, because of the acidic character of the aromatic hydroxy group. However, hydroxyethoxylated bisphenols can be used with advantage because of the bulky rigidity of the aromatic ring and the isolation of the terminal hydroxyl groups from the π electron system.

All of these reactions can be telescoped by mixing all ingredients together simultaneously in a one shot process in which event the groupings in the chain are randomly arranged. Whichever system is followed, it is essential to drive the reaction to completion by post heating to above 100°C despite the thermal assistance provided by the larger reaction exotherm in the case of the one shot system.

At this temperature a certain amount of further reaction occurs with the excess diisocyanate which reacts with the active hydrogen carried by the already formed urethane groups (if glycol chain extended or urea groups if diamine extended) to form allophanate and biuret crosslinks respectively. These may terminate on other polymer chains or in a triazane ring formed from some trimerised diisocyanate. A large excess of diisocyanate increases the chemical crosslink density and hence the hardness of the final polymer, although the low temperature properties and elongation at break are affected progressively more and more.

The allophanate and biuret linkages are less thermally stable than the urethane linkages and therefore break down at injection moulding temperatures, reforming again on cooling. Superimposed on this purely chemical crosslinking effect is the formation of crystalline regions or 'knots' following alignment and hydrogen bonding of the short chain glycol diisocyanate segments which are interleaved with flexible soft domains formed from the long chain polyether or polyester and diisocyanate. The presence of these 'knots' acts to prevent creep and adds to the temperature range over which softening takes place. This in turn facilitates easy moulding and minimised drool. The extent of hydrogen bonding and hence recrystallisation depends on the time and mobility of the flexible segments and hence on the rate of cooling. Stable super-cooling is readily achieved by rapid chilling. The transparency and stiffness are correspondingly affected.

When this chemistry is referred to the coating field, it is found that whilst highly chemically crosslinked urethane is insoluble in almost all solvents, the thermoplastic urethanes based on hydrogen bond crosslinks dissolve in aprotic solvents such as dimethyl formamide, tetramethyl urea, dimethyl suphoxide, N methylpyrollidone, etc., largely by the breaking up of hydrogen bonding forces existing between the active urethane hydrogen atom on the urethane nitrogen with the carbonyl oxygen on an adjacent urethane group. Accordingly for this application it is desirable to minimise the number of chemical allophanate and biuret crosslinks and yet at the same time maintain an adequate M.W. to ensure reasonable mechanical properties. This can be achieved by limiting the final hot curing of the urethane and stabilising the polymer against further crosslinking when put into solution.

Optimum properties depend on using very pure MDI and very regular polyesters or polyethers to ensure maximum hydrogen bonding between the hard segments and good packing between the flexible segments.

When using polyethers, the solubility problem is eased by the attraction of the repeat ether oxygen atoms for protons and therefore these polymers disperse to give relatively highly concentrated solutions in alcohol/ether/ester/ solvents. However, where maximum dry cleaning resistance, gas impermeability, etc., is required, the polyester system is preferred.

An alternative technique which is really outside the scope of this paper is the use of masked isocyanates added to the solution which dissociate to release diisocyanate to form chemical crosslinks when the coating is baked after driving off the solvent.

Certain diisocyanate dimers, particularly TDI dimer, aminimides, HCN blocked diisocyanates can be used as they all decompose in the range 80°–150°C.

a)

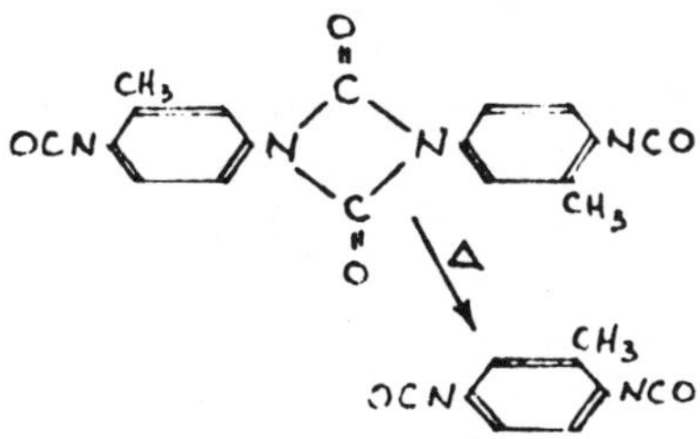

TDI dimer

b)

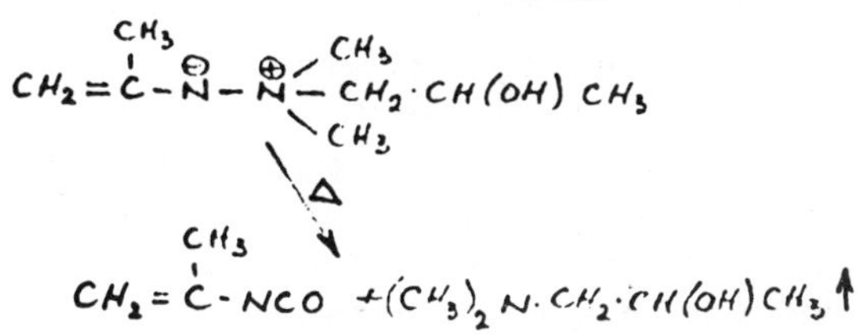

Dimethyl 2-hydroxypropyl-ammonium N-methacryloyl-amine (1,1 dimethyl 1-(2 hydroxypropyl) amine-methacryl-imide)

PRACTICAL FORMULATION

Urethanes prepared as described above are available in the form of chips or extruded pellets in grades for use by moulding or in grades for use as coatings after dissolving in suitable solvents. Ready prepared one component solvent solutions varying in solids content from 25 to 60% are also available commercially. Incorporated into the better quality materials are minor additions of auxilliary chemicals which typically comprise:

1. Anti oxidants – usually phenothiazine or hindered phenols.
2. Ultraviolet light absorbers – usually benzimidazoles, benztriazoles or benzophenones.
3. Anti hydrolysis agents – usually carbodiimides.
4. Figments, fillers, flatting agents.
5. Lubricants and plasticisers.

Of the above, some comment must be made first about anti oxidants, lubricants and U-V absorbers.

It is apparent that various trace metals can end up in the urethane viz:

(a) Sodium and potassium from polyethers and polycaprolactone.
(b) Iron from diisocyanates arising from HCl corrosion of phosgenation equipment.
(c) Tin, cobalt, lead, zinc, etc., from esterification catalysts used to shorten esterification time when preparing polyesters or contained in metal soap lubricants.

Many of the metals, even in trace quantities, markedly affect the rate and course of the various competing reactions into which diisocyanates can enter. Finally, transition metals are known to catalyse aerial oxidation of hydrocarbon sequences in all polymers, an effect enhanced by singlet oxygen arising from U-V radiation in the presence of air. The resultant peroxides readily hydrolyse leading to chain scission. Additionally by proton extraction from aromatic alkyl side chains coloured products are formed, and this accounts for the yellow-browning of urethanes based on aromatic diisocyanates.

Again as polyesterification is a reversible reaction whose rate but not equilibrium is affected by metallic catalysts hydrolytic scission of the polyester chains sequences more quickly occurs in their presence.

In so far as polyethers are hydrophilic, the higher water content arising from their presence if used in conjunction with polyesters is disadvantageous because although the ether linkage is stable to hydrolysis the ester linkage is not.

Furthermore, when the acid element of the ester is a strong acid the initial liberation of this acid acts as a catalyst to further hydrolysis. Only in the case of the carbonate linkage is this effect absent because carbonic acid is such a weak and unstable acid. In consequence, only polycarbonate diols chain extended with glycols and diisocyanates are truly stable to hydrolysis. The urethane linkage, although an isocyanic acid ester, appears to be moderately resistant to hydrolysis, possibly due to the ultimate amine product scavenging any strong acid present.

To minimise hydrolytic degradation it is, therefore, essential to add a material with a long life and continuously reactive to water and strong acids forming innocuous products. Ketenes and carbodiimides are an obvious choice, particularly the latter as it is so readily formed by catalytic decarboxylation of isocyanates:

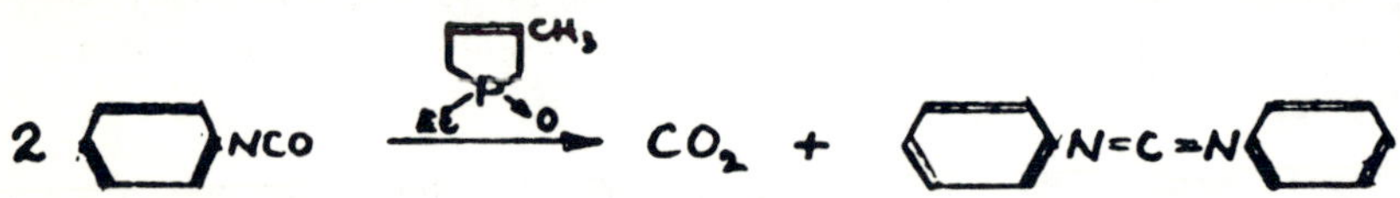

Diphenylcarbodiimide

Such simple carbodiimides react too rapidly with water and have too short a life and, therefore, sterically hindered carbodiimides are used, typically Di(diisopropylphenyl) carbodiimide (STABAXOL I – (BAYER)).

Because this material has limited solubility in urethane and tends to bloom at the surface, the corresponding polymeric carbodiimide is also available (STABA-MOL PCD (BAYER)). This material can effectively re-link carboxylic chain ends.

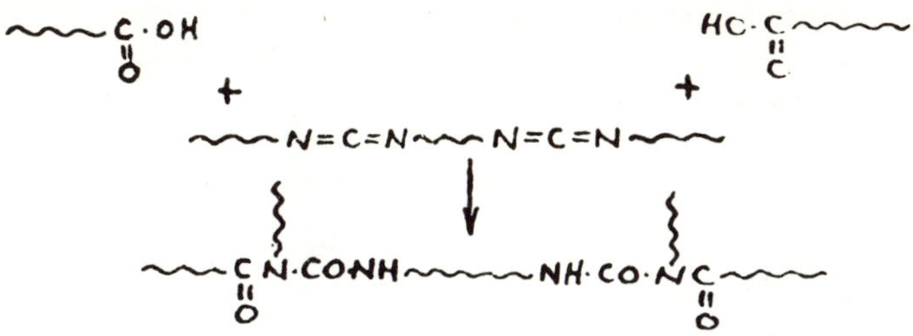

By judicious compounding with the above materials a practical moulding or coating material can be produced with good to excellent resistance to degradation by sunlight and hot, humid conditions.

PHYSICAL PROPERTIES

Thermoplastic urethanes are block copolymers in the same sense as other well known hydrocarbon block copolymers, such as ABS. The special requirements are the same, i.e., a hard glassy homopolymer component domain chemically linked to a soft rubbery homopolymer component domain. In the case of ABS the former is polystyrene, PMMA, etc., whilst in the thermoplastic urethane it is the high melting point urethane resulting from reaction of the diisocyanate with a short chain glycol or diamine. The high melting point and associated glass transition temperature well above room temperature arise from the rigidity of the aromatic isocyanate ring coupled with the high density of hydrogen bonds between the amido protons and the carbonyl oxygen. Control of block M.W. and distribution is readily and easily effected by careful selection of the polyester or polyether and the amount of diisocyanate and chain extender used.

In so far as the two homopolymers must have M.W.'s in excess of about 2500 for the hard segment, and about 10,000 for the soft segment to ensure mutual insolubility and hence the formation of hard and soft domains, the former will have a minimum size of about 400 Å (if considered spherical). Electron microscopy shows that the hard domains are not spherical, but rod like with lengths of 3000–8000 Å and widths of about 25 Å.

The existence of these domains is also clearly demonstrated by changes in optical bi-refringence and the depolarisation of scattered light with variation of both stress and strain. The stress and strain optical relationships are quite different, and this differentiates block copolymers from simple rubbers.

The hard and soft segments appear to be arranged in layers as lamellae. In so far as the average M.W. of the urethane chain is about 50,000 it is clear that the same molecule is threaded through and folded within both hard domains and the amorphous soft domains formed by the association of the various polyester (or polyether) segments. These latter regions only show evidence of crystallinity when the polymer is stretched, whereas the hard domains show evidence of at least 75% crystallinity (hydrogen bonding) at all times. The separation into domains is more evident with polyester than with polyether type urethanes and longer polyester segments lead to enhanced domain structure. Probably hydrogen bonding again differentiates the polyester from the polyether.

MECHANICAL PROPERTIES

The hard glassy domains act as both a reinforcing filler, and as a tie point between adjacent rubbery soft segments, and which, therefore, interfere with chain unravelling which would otherwise tend to occur when subjected to stress.

The mechanical behaviour of thermoplastic urethane is characteristic of a block copolymer, and other mixed systems in that the plot of Modulus against temperature shows two glass transition temperatures, relating to the two homopolymers, linked by a region of enhanced rubbery modulus. Although a decrease in modulus by a factor of perhaps, 100 occurs at the Tg of the soft segment, an unusually high modulus is maintained up to the Tg of the hard segment. The extent of this transition region is moreover effected by the stress history of the sample.

By contrast, a simple co-polymer exhibits no platean arising from independent Tg's of two phases, but as the composition is varied the single Tg varies proportionately, and the change in modulus in traversing this Tg region is frequently of the order of 1000, i.e., 10 times that for a two phase system.

In the case of thermoplastic urethanes a change in the type of diisocyanate used affects mainly the level of platean modulus (more bulky diisocyanates yielding a higher modulus, presumably due to the increased filler effect). Changes in the polyester composition effect only the lower Tg and not the platean modulus. This lower Tg is increased as the frequency of the ester linkages increase, presumably due to reduced flexibility arising from increased van der Waals forces. The effect of the chain extender used is broadly similar to that of the diisocyanate, i.e., bulky aromatic glycols raise the platean modulus more than aliphatic glycols.

Plasticiser inclusion varies according to whether the plasticiser affects most the soft rubbery segment or the hard crystalline segment. Chlorinated aromatics, such as polychlorinated biphenyl lower the modulus platean whilst low M.W. aliphatic polyester plasticisers reduce the Tg of the rubbery phase and hence the low temperature stiffening point of the urethane without having any major effect on the platean modulus. Additional crosslinking has an effect again depending on whether it is located in the soft or hard segment. In the case of ABS type polymers crosslinking in the hard segment is chemically difficult and usually sulphur or peroxide vulcanising occurs in the soft rubbery segment. By contrast in the case of urethane, the low reactivity of the soft polyester or polyether segment makes it difficult to form crosslinks in this region. However, crosslinks in the hard crystalline region are readily accomplished by the use of excess diisocyanate which forms allophanate, biuret or triisocyanurate crosslinks, or by the inclusion of triols, quadrols, etc., in part replacement of the glycol chain extender.

Increase of excess diisocyanate increases both the platean modulus and the modulus above the upper Tg (i.e., melt viscosity) whilst the use of polyhydroxy chain extenders lowers the platean modulus, but still increases the modulus above the upper Tg. This effect is considered to be due to the disordering effect of the polyhydroxy compound on the ordered diol/diisocyanate crystalline structure, as well as by introducing crosslinks. In terms of coatings technology the effect is, of course, detrimental in that solubility (or solvent swell) decreases whilst tensile modulus decreases as the Molecular weight per crosslink decreases.

If the strained but unbroken block copolymer is relaxed to zero stress some stress softening is observed in that on re-straining a different stress/strain path is followed, and the material appearing to have reduced modulus, whilst after about 5 cycles the stress softening effect levels out. In the case of thermoplastic urethanes which have less than 30% of hard glassy domain homopolymer, this effect of stress softening is not observed.

The original properties may be recovered by annealing at elevated temperatures for several hours (say about 150°C) or by melting (re-moulding).

In the case of films cast from solvents, the choice of solvent also has an effect on the mechanical properties; if the solvent preferentially dissolves the soft segment, the hard glassy phase is precipitated on evaporation to form a more or less continuous phase surrounding pockets of soft segment material. The result is a considerable fall in initial modulus.

Again, when the proportion of hard glassy domain material is less than about 30% this effect is not very evident, and choice of solvent is not so critical.

Unfortunately, due to the amount of work involved very little stress relaxation data has been accumulated on urethane block copolymers to enable the construction of reliable master curves via time-temperature superposition. Data so far available indicates that the Williams-Landel-Ferry relationship is not followed in the rubbery platean region. As mentioned earlier, this is evidenced by opto-mechanical observations on stressed specimens and demonstrates the presence of multiple relaxation processes.

When dissolved in solvents the urethane block copolymers display some unusual effects depending on concentration – concentrated solutions show evidence of ordering not characteristic of the parent homopolymers, whereas dilute solutions are more Newtonian in behaviour such that viscometry, and osmomometry of these dilute solutions provide meaningful data. With solvents which predominately dissolve one phase, the other phase tends to concentrate into separate regions of the solvents. If the solvent is a mixture whose composition varies with evaporation, this effect can be increased and the cast film made more or less inhomogeneous resulting in the limit in a variation in porosity. This effect is exploited in the production of moisture permable, breathable, coatings for simulated leather products, leather lacquers and the like where the microscopic variation in composition and concentration lead to a type of microporosity or reticulation.

In the melt the two phase block copolymer nature of the thermoplastic urethanes result in anomalous viscosity effects. Compared to continuous phase copolymers whose melt viscosity varies in general with molecular weight to the power of 3, 4 at higher molecular weight levels the melt viscosity of 2 phase co-polymers varies as the 5.5 power (at low shear rates). This may be interpreted as the retention of some residue of 2 phase structure in the melt. The effect is also shown in dynamic viscosity data which at low frequencies display a marked elastic response whilst at high frequency the melt exhibits a response characteristic of single phase thermoplastic materials. In practical terms, erratic melt swell and rough surface finish can easily result from extrusion at too high a rate or at too low a temperature.

REFERENCES

1. Bonart, V. R. – J. Macromol. Sci – Phys B 2(1) 115–138 (1968).
2. Bonart, V. R.; Morbitzer, L; Rinke, H. – Kolloid-Z u Z. Polymere *240* 807–819 (1970).
3. Campbell, T. W.; Smeltz, K. C. – J. Org. Chem. *28* 2069–2075 (1963).
4. Cella, R. J. – J. Polymer Sci. Symp. *42* 817–832 (1973).
5. Clough, S. B.; Schneider, N. S. – J. Macromol. Sci., Phys B 2 (4) 553–566 (1968).
6. Ehrenberg, E. G., Piskarava, E. P.; Poddubnyi, I. Ya. – J. Polymer Sci. Symp. *42*, 1021–1030 (1973).
7. Estes, G. M.; Cooper, S. L.; Tobolski, A. V. – J. Macromol. Sci. – Revs. Macromol. Chem. C4(2) 316–366 (1970).
8. Frisch, K. C. ed. – Cyclic Monomers (High Polymer Series XXVI) (1972).
9. Frisch, K. C. Reegan, S. L. ed. – Ring Opening Polymerisation (1969).
10. Frisch, K. C.; Reegan, S. L.; Ulrich, H.; Schollenberg, C; Singh, A. – Chemistry and Physics of Coatings and Films – Gordon Conference, Tilton N-H., U.S.A. (1970).
11. Gable, C. L. – Polyurethane Elastoplastics – Appl. Polyer Symp. in New Polymeric Materials Ed. P.F. Bruins *11* (1969).
12. Hespe, V. H.; Meisert, E.; Eisele, U.; Morbitzer, L.; Goyart, W. – Kolloid-Z u Z. Polymere *250* 790–811 (1972).
13. Kogan, I.C. – J. Org. Chem. *26* 3004–5 (1961).
14. Macknight, W. J.; Yang, M. – J. Polym. Sci. Symp. *42* 817–832 (1973).
15. Magnus, G.; Nunleary, R. A.; Critchfield, F. E. – Rubber Chemistry and Technology (Sept. 1966).
16. Meerwein, H.; Delfs, D.; Marschel, H. – Agnew. Chemie. *72* 927 (1960).
17. Mieras, H. J. M. A. – J. Polym. Sci. Symp. *42* 987–1000 (1973).
18. Millar, G. W. – J. Appl. Polym. Sci. *15* 39–45 (1969).
19. Millar, G. W.; Saunders, J. H. – J. Appl. Polym. Sci. *13* 1277–1288 (1969).
20. Neumann, W.; Peters, J.; Holtschmidt, H.; Kallert, W. – Preprints, 4th Rubber Technology Conference, London (May 1962).
21. Nishi, T. – Appl. Polym. Symp. – U. S./Japan Seminar on Polymer Processing and Rheology *20* 353–363 (1973).
22. Okeda, H.; Iwakura, Y. – Macromol. Chem. *64* 91–101 (1963).
23. Ossefort, Z. T.; Testroet, F. B. – Rubber Chem. Technol. *39* 1308 (1966).
24. Piggott, K. A.; Britain, J. W.; Archer, W.; Frye, B. F.; Cote, R. J.; Saunders, J. H. – Ind. Eng. Chem. Prod. Res. Dev. *1* 28 (1962).
25. Smith, H. A. – J. Polymer Sci. A1(6) 1299 (1968).

W. D. DAVIES

Principal Scientist
Central Research and Development Dept.
Lankro Chemicals, Ltd.

URETHANE COATED TEXTILES*

(Received December 27, 1974)

ABSTRACT: This article gives a detailed discussion of urethane coating textiles using the transfer process. One-component systems have overcome many earlier disadvantages by requiring only the evaporation of the solvent to achieve a urethane film with fully developed physical properties. Tests are described for determining hydrolysis resistance, abrasion, flexing, and adhesion.

KEY WORDS: Urethane textiles, transfer coating, hydrolysis resistance, fabric testing.

INTRODUCTION

TEXTILE COATING TODAY is a very large and diverse industry and one of the areas that has expanded most rapidly in the last 5 years or so is that of urethane coated fabrics. Like oilcloth, nitrocellulose, PVC and expanded PVC before it, urethane is a coating designed to try and imitate leather. Hence the products find application as a general substitute for leather in the upholstery, garment, footwear, luggage and allied industries.

Urethane based coatings have been used for many years for application to hard surfaces as protective films and lacquers where their excellent weathering properties, toughness and abrasion resistance have been extremely beneficial. More recently urethanes possessing more extensible and elastic properties have been used with success for application to woven fabrics such as nylon, to produce light-weight waterproofing materials, or to special raised-pile woven fabrics to yield a product with the appearance and certain properties of natural leather.

Transfer coating and direct coating are the most common methods of application of urethanes to textiles. In both methods, the urethane is applied from a solution.

This contribution will be devoted to a detailed discussion of the various aspects of the transfer process.

*This paper originally presented at the Polyurethanes Symposium, University of Surrey, Guildford, Surrey, England, September 23–25, 1974, organized by the Polymers Paint & Colour Journal.

TRANSFER COATING

This technique which was originally designed for the PVC coating industry is particularly suitable for adaptation to coating materials with urethanes since they must of economic necessity be applied in relatively thin films, usually less than 0.004″ dry film thickness. An advantage of perhaps even greater importance associated with the adoption of the transfer process is the ability to apply urethane coatings to a considerably wide range of substrates which can include extremely light-weight brushed nylon, woven raised pile fabrics, knitted fabrics, non-woven fabrics and natural skins such as sheepskin and leather.

The transfer coating process for urethanes was initially developed in the mid-1960's though it was not until 2–3 years ago that coated fabrics first gained commercial acceptance in this country.

This technique involves the application of a urethane solution normally referred to as the top coat to a suitably treated paper surface, drying or curing this urethane film by means of heat, applying a second urethane coating known as the adhesive or tie-coat and whilst this second coat is in an adhesive state, applying the required substrate under controlled pressure.

The sandwich or laminate so formed is passed through a second heating stage and is then cooled and the paper removed from the urethane coated substrate by winding the two parts onto separate spools.

A diagrammatic representation of a modern plant for producing urethane coated fabric by the transfer process is given in Figure 1.

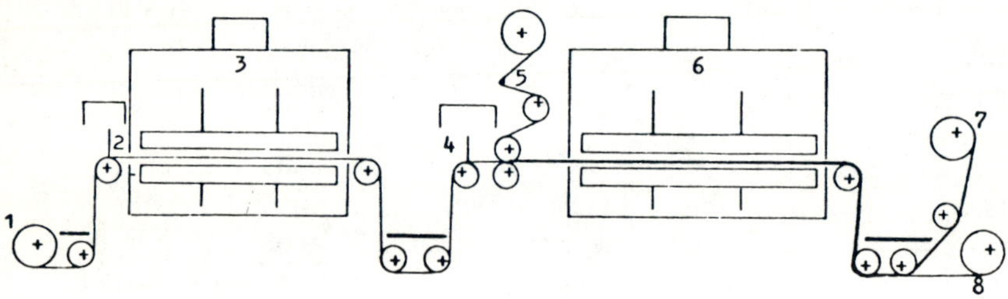

1. RELEASE PAPER UNWIND
2. FIRST COATING HEAD
3. FIRST ZONED OVEN
4. SECOND COATING HEAD
5. FABRIC LAMINATION UNIT
6. SECOND ZONED OVEN
7. COATED FABRIC UPWIND
8. RELEASE PAPER UPWIND

Figure 1. Transfer coating unit.

In the early days of transfer coating most of the machines comprised a single coating head which meant that after the application of the top coat the paper had to be transferred from one end of the coating machine to another.

A further feature of the process present in those days but which has now disappeared was the use of two-component systems for both top and tie-coats. By their nature, two-component systems are rather soft and slightly tacky until full cure is achieved, hence one of the first problems encountered was that a 'double-sided' release paper had to be used. This type of release paper was necessary with single head coating machines since the paper plus the first coat had to be rolled up before applying the second coat.

As a consequence of the relatively long cure time the coated fabric cannot be stripped off the paper immediately since there would be a tendency for any loose fibres on the back of the fabric to be picked up by the tacky top coat. Furthermore since the coating has to be left in contact with the paper for a period of up to 24 hours to allow cure to take place, there is no immediate opportunity to correct any faults.

In view of these major disadvantages accompanying the use of two-component coating systems for the first coat, attention turned to the use of one-component coating systems requiring only the evaporation of the solvent to achieve a urethane film with fully developed physical properties. The choice between a one or two-component tie-coat is still open to the coater and has resulted in many heated arguments over the last 2–3 years. This question will be discussed later in the paper.

PROCESSING REQUIREMENTS

At this point it is as well to list the four main items that a coater has to consider before commencing production of coated fabrics: (1) Urethane solution, (2) Release Paper, (3) Coating Machine, (4) Fabric.

The last two items are outside the scope of this paper and will only be discussed where relevant.

Regarding the urethane solutions and the release paper, it should be stressed that the correct combination of these two items is vital for the production of a technically sound coated fabric.

The release paper selected should be suitable for use with urethane polymer solutions. Such papers are designated by the manufacturers as 'Urethane grade' to distinguish them from papers designed for transfer coating PVC based compositions, though in certain instances urethane solutions may be transferred from the latter grade.

The release characteristics of different types of urethane and PVC grades of paper can vary from 'tight' release to 'easy' release. The relative ease of release plays a significant part in determining the correct combination of top coat and tie-coat.

In order to decide upon the correct combination of top and tie-coat, one must have some knowledge of the solubility characteristics of the urethane polymers. Most of the urethane top coats that have been used to date consist of 25–30% solutions of fully reacted polyesterurethane thermoplastics based on aromatic isocyanates. Dimethylformamide (DMF) is the most widely used solvent for these polymers but due to its high boiling point, cost and toxicity, it is usually used in combination with diluents such as methyl ethyl ketone (MEK) and toluene.

For a typical urethane top coat the relationship between the solution viscosity and solvent composition at a fixed solids content is represented by the graph in Figure 2.

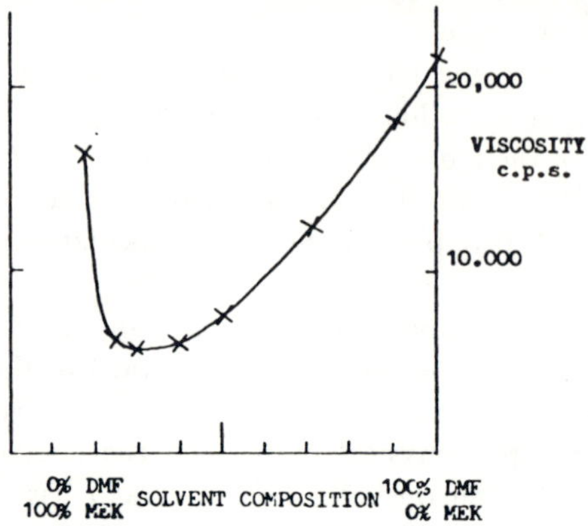

Figure 2. Viscosity vs. solvent ratio.

Using such a graph some predictions can be made as to how a film of such a polymer would behave when tie-coats containing different solvent composition are applied to it. First of all, let us consider the two extreme cases, namely a tie-coat in 100% DMF and 100% MEK respectively. In the first instance, we can deduce that such a tie-coat will tend to redissolve the top coat and hence at the lamination stage the fabric pile will penetrate right through both coats leading to an undesirable product.

In the second case the solvent is too weak to dissolve the top coat and its only action is to swell it thus giving rise to the phenomenon of 'crazing'. The extent of crazing will obviously depend on how much the top coat actually lifts off the paper, that is, on the ease of release.

In practice, however, one seeks the conditions whereby the top coat is only sufficiently attacked to allow some penetration of fabric pile. These conditions usually coincide within the region of minimum viscosity in the graph. Hence from a graph of this type one can obtain an approximate set of conditions. Since this curve represents static conditions, the optimum solvent balance for transfer coating can only be found by carrying out actual coating trials using tie-coats containing different solvent ratios and different release papers.

To illustrate this point, coatings were produced on two grades of release papers with widely different release properties. The tie-coats used consisted of 28% solutions of a urethane polymer in a variety of solvent compositions. A summary of the results is given in Table 1.

In the early days crazing was overcome by addition of DMF to the tie-coat but with little regard to the effect it had on the performance of the coated fabric. This was particularly so at the beginning of the 'wet-look' era when most of the high

Table 1.

SOLVENT COMPOSITION		RELEASE PAPER			
		BINDA MB 40(EASY RELEASE)		BINDA 008(TIGHT RELEASE)	
% MEK	% DMF	APPEARANCE	FLEX PERFORMANCE	APPEARANCE	FLEX PERFORMANCE
30	70	No crazing	0.5 (G2) 1.0 (G2)	No crazing	0.5 (G2) 0.5 (G2)
40	60	Slightly crazed in isolated areas.	1.0 (G2) 1.0 (G2)	No crazing	0.5 (G1) 1.0 (G2)
50	50	Slightly crazed in isolated areas.	0.5 (GO) 1.0 (G2)	No crazing	1.0 (G2) 1.0 (G2)
60	40	Extensively crazed in isolated areas.	0.5 (GO) 1.0 (GO)	No crazing	1.0 (GO) 1.0 (GO)
70	30	Severely crazed -crocodile skin effect.	0.5 (GO) 1.0 (GO)	No crazing	1.0 (GO) 1.0 (GO)

NOTE: (1) Release papers supplied by CARTIERE AMBROGIO BINDA S.p.A, MILAN.

(2) The flex performance was determined using a Schildknecht flex tester. The samples were flexed to 0.5 x 10^6 and 1.0 x 10^6 respectively and the extent of damage was assessed.

(3) The notation in brackets refers to the extent of damage after flexing. Grade 0 represents no damage ; Grade 4 represents extensive damage.

gloss fabrics were achieved by transfer off a high gloss release paper. In order to minimise the inevitable tendency of coated fabrics with a patent finish to 'cling', silicone oils and other 'slip aids' were added to the top coat. The result of this was that the release from the paper was too easy hence the need for additional DMF to overcome crazing.

So far we have discussed the importance of controlling the solvent composition of the tie-coat. Next it is important to consider the control of "pile penetration" into the top coat. The inability to control the degree of pile penetration has a serious effect on the flex performance of the coated fabric as shown in Table 1.

These results support the theory that as the degree of pile penetration increases with increasing DMF content of the tie-coat, so does the extent of damage after flexing.

We must now consider in more detail the factors that influence the extent of fabric pile penetration during the transfer process.

For a conventional urethane system the following points are considered to be of prime importance.

(1) Nature of top coat.
(2) Thickness of top coat.
(3) Solvent composition of tie-coat.
(4) Solids content of tie-coat.
(5) Time delay between application of tie coat and contact with the fabric.
(6) Laminating pressure and fabric quality.

The time delay between application of the tie-coat and the contact with the fabric is important since the greater the time lapse, the greater the degree of attack will be.

Hence we can see that this parameter will depend on the facilities provided on the coating machine as well as the operating speed. The distance between the second coating knife and the laminator can vary between six and sixty inches whilst the operating speeds are usually of the order of 5–10 yards/minute. This means that the time delay between the application of the tie-coat and contact with the fabric can vary between 1 and 40 seconds. It is not surprising therefore that coatings produced under these extreme conditions exhibit different physical properties. Bench coating experiments illustrating these points appear in Table 2.

Table 2.

Tie-coat		Schildknecht flex performance(x 10^6).			
% MEK	% DMF	3 second delay		60 second delay	
100	-	1.9	(GO)	No coating produced. Top coat lifted off the paper.	
85	15	1.9 1.9	(GO) (GO)	1.9	(GO)
70	30	0.95 0.43	(G3) (G1)	0.46 0.46	(G3) (G2)

From these results it can be seen that the flex performance deteriorates firstly through increasing the DMF content in the tie-coat, and secondly, by prolonging the delay time.

In the case of the tie-coat based on 100% MEK and a delay time of 60 seconds, the top coat crazed to such an extent that it lifted off the paper completely. Ideally, the rate of attack on the polymer in the top-coat should be such that it can tolerate as wide a range of coating conditions as possible. It should be realised however that the rate of attack on the top coat, by the solvents in the tie-coat, will depend on the pigments and fillers employed as well as the structure of the urethane polymer. Finally the extent of pile penetration will depend on the laminating pressure and the fabric pile quality. Higher laminating pressure and an irregular long pile will tend to produce coated fabrics with inferior flex performances.

To a large extent, the 'break' of the coated fabric will depend on the fabric pile density and the laminating pressure. Hence it can be seen that in order to produce a coated fabric with both good physical properties and appearance, careful control must be exercised on all parameters discussed above.

Most of the faults that occur in urethane coated fabrics can usually be attributed to the lack of control exercised over one or more of the parameters discussed above.

TEST METHODS

At this point we must consider the performance of the coated fabrics. Since urethane coated fabrics are still in their infancy the test methods set up to assess their performances are only tentative.

It is now proposed to discuss the testing methods currently used in the Industry and the relevance of such tests.

Most of the work in this field has been carried out by the Furniture Industry Research Association (FIRA) and the Shoe and Allied Trades Research Association (SATRA).

Initially, urethane coated fabrics received a great deal of adverse publicity due to their poor performance in various end applications. This coincided with an apparent low resistance to breakdown on laboratory accelerated hydrolysis tests. The two factors therefore tended to be related and hence hydrolysis resistance became identified with good performance.

Some of the more important test methods currently being employed in the coating industry will now be discussed.

Flex Performance of Urethane Coated Fabrics

The flex performance of a urethane coated fabric is a very important parameter since with few exceptions the end application for such a material means that it will be subjected to varying degrees and modes of flexing. The extent and mode of flexing can vary tremendously from one application to another and this in turn means that there will be appropriate specifications laid down by the manufacturers of the end product. These specifications may be arbitrary or they may be the results of extensive correlation tests between 'in wear' and accelerated flex testing.

A number of specifications have emerged which state the minimum number of flex cycles a urethane coated fabric should survive without cracking on a particular flex tester. A summary of the specifications, currently in use follows:

Application	Flex tester	Specification	Comments
Footwear	SATRA Vamp	1 million	Widely used in the U.K.
Footwear	BALLY	100,000-150,000	Widely used in other European Countries.
Upholstery	Schildknecht	400,000	FIRA specification
Garments	Any of the above.	Not as well defined.	

Although most manufacturers of urethane coated fabrics supply to the industries concerned with all three of the main applications listed above, very rarely do they possess more than one type of flex tester. Hence it is usual for the coated fabric manufacturer to set his own standard on one flex tester. This immediately presents the problem of correlating the performances of the different types of flex testers. Extreme care must be taken when carrying out such a correlation, assuming one exists at all. As a general guide we believe the following correlation to hold.

$$\text{Bally } (100{,}000) \simeq \text{Schildknecht } (1 \text{ million}) \simeq \text{Satra Vamp } (1 \text{ million})$$

The extent of cracking is graded on an arbitrary scale ranging from 0 to 4, Grade 0 representing no damage and Grade 4, extensive damage.

Since the majority of the faults that have been encountered in the coating industry showed up in the form of cracking, then obviously the assessment of the flex performance of a urethane coated fabric both before and after accelerated aging tests is vital in attempting to predict the life time of such materials. Currently 400,000 cycles on the Schildknecht tester is considered acceptable by F.I.R.A. but 1,000,000 cycles without damage is readily attainable.

Adhesion

Due to the low strength and highly elastic character of the urethane top coat, the adhesion of a urethane coated fabric cannot be measured directly. The method of measuring adhesion most widely used involves an adaptation of that recommended in B.S.3424 (Method 9).

With the exception of coatings with very low adhesion, it is usual for the pile to pull away from the fabric – hence in many ways this test is a measure of the pile strength.

In certain instances the directional influence of the pile can be observed since differences in peel bond strengths can be obtained depending whether delamination is in the same or the opposite direction to that of the pile. Some results illustrating this point are recorded in Table 3.

Table 3. Influence of Coating Conditions on Adhesion

Tie-Coat	COATED WITH PILE DIRECTION.			COATED AGAINST PILE DIRECTION.		
	DIRECTION OF ADHESION TESTING.			DIRECTION OF ADHESION TESTING.		
	Weft	WARP		Weft	WARP	
		With Pile	Against Pile.		With Pile	Against Pile
T1 one - component	(1) 3.4 (2) .3.4	(1) 3.0 (2) 2.8	(1) 5.6 (2) 5.3	(1) 3.5 (2) 3.3	(1) 2.5 (2) 2.6	(1) 4.5 (2) 4.2
T2 two - component	(1) 2.7	(1) 2.3 (2) 2.4	(1) 6.0 (2) 6.0	-	-	-
T3 two - component	(1) 2.3 (2) 2.4	(1) 1.8 (2) 1.6	(1) 7.5 (2) 7.2	(1) 2.5 (2) 2.5	(1) 1.9 (2) 2.2	(1) 6.7 (2) 7.1

NOTE: Results are expressed as Kg/5 cm.

From these results it can also be seen that the direction of the pile relative to the direction of coating does not seem to have had much bearing on the adhesion. On a full scale transfer coating plant, a great deal of importance is placed on ensuring that the fabric pile is presented in the correct manner at the laminating head.

The most common method of ensuring that the pile is not too flat at the laminating head relies on the geometry of the coating machine. Other methods involve brushing or passing the fabric through an electrostatic field in order to achieve consistent results.

The strength of the bond between the top coat and the fabric depends on a number of factors, including the extent of pile penetration into the top coat; the adhesive strength of tie-coat; the viscosity and solids content of tie-coat and the tear strength of the fabric pile.

The greater the degree of pile penetration, the greater the bond strength will be. Hence the adhesion will be dependent upon the solvent balance of the tie-coat. This is more critical in one component tie-coats than in two component tie-coats, since the latter are better adhesives in their own right and do not rely on pile penetration to the same extent. This is shown in Table 4.

Table 4.

	TIE COAT COMPOSITION		ADHESION (Kg/5 cms).WARP DIRECTION	
	% MEK	% DMF	With Pile	Against Pile
One Component Tie-Coat	70	30	11.4	9.7
	80	20	6.7	5.4
	85	15	5.0	4.7
	90	10	4.7	4.7
Two Component Tie-Coat	90	10	5.6	6.8
	80	20	4.7	4.7
	70	30	4.9	4.6
	60	40	4.6	6.0
	50	50	5.4	7.0

These results show clearly the dependence of adhesion on the solvent composition of one component tie-coats. In contrast two component tie-coats are essentially independent of the tie-coat solvent composition.

We conclude therefore that for one component tie-coats in particular, the factors that promote adhesion adversely affect the flex performance. Coated fabrics with low adhesion will tend to delaminate in use thus detracting from their visual appearance and at the same time rendering them more vulnerable to 'picking' by sharp objects.

Scuff Resistance

Doubts have always been expressed about the scuff resistance of urethane coated fabrics due to the thin coatings used. The most commonly employed test used for assessing the scuff resistance is that proposed by FIRA. The test involves abrading specimens against various grades of emery paper using a Martindale tester. The results from this test vary considerably from one testing machine to another. The scuff resistance is dependent on the base fabric, the adhesion and the actual weight of coating as well as the coating uniformity. The last two factors are the most important. In the following series of coated fabrics, the abrasion resistance is seen to be dependent on the coating weight.

Ref No	Scuff rating	Coating weight. g.s.m.
VP 9	Grade 0	59
VP 10	Grade 1	69
VP 11	Grade 3	77

Appearance of Coated Fabrics

Examination of coated fabrics under a microscope can reveal evidence of defects such as pile penetration, pinholing and blistering.

Pinholes arise as a result of inadequate control over the solvent removal from the top coats. As the importance of the solvent balance in the tie-coat has been discussed earlier, so now we must discuss the solvent balance of the top coat.

In order to produce a top coat free of pinholes and blisters, correct balance between the amount of high and low boiling solvents must be maintained. A further requirement is that efficient ovens with good temperature control are used.

Pinholes may also arise due to air being incorporated during pigmentation, hence all top coats should be deaerated prior to coating.

A 'pinholed' coated fabric will pick up dirt which will remain in the pinholes giving rise to a 'microdot' printed appearance even after attempts to clean the fabric. Although this type of fault can occur in any urethane coated fabric, it will be much more apparent in lighter coloured materials.

Hydrolysis Testing

Of all the aging tests that have been carried out on urethane coated fabrics, hydrolysis testing has been by far the most popular, and is most probably the most controversial.

A summary of the accelerated hydrolysis tests that have been employed in the coating industry, are given in Table 5.

It is necessary at this juncture to analyse the test conditions critically. The chemical reactions that are evident at the high temperatures used in the tests may not necessarily take place under normal wear conditions and these temperatures may in some instances approach the softening point of the polymer. Results also lack consistency and anomalous results can be obtained, for instance the flex performance of a coated fabric may be improved after hydrolysis.

Table 5. Hydrolysis Resistance Tests

1.	Boiling in 4% sodium hydroxide solution for 15 minutes.
2.	Storing for 1 week at 70°C and 75% relative humidity.
3.	Storing for 1 week at 70°C and 95% relative humidity.
4.	Storing for 8 weeks at 70°C and 95% relative humidity.
5.	Heating for 72 hours in water at 70°C and 90°C.
6.	Bayer's Tropical test - storing for 7 and 14 days at 70°C and 95 - 100% relative humidity.
7.	FIRA hydrolysis test - 16 hours in steam at 100°C.
8.	SATRA CM 28 test method - 15 hours in urea solution (1.2 g/l)at 100°C and 120°C.
9.	A 'solubility' test (Ref 1)
10.	Ford test - 6 hours in steam at 130°C

Of the test methods currently in use, the so called "tropical test" seems to be the most realistic, whereas the "solubility" test appears the least realistic.

Most of the elastomers that are used for coating fabrics are polyester based urethanes. Since esterification is a reversible process then it is only to be expected that these elastomers will break down on hydrolysis. Several excellent papers have been published on the subject of the hydrolysis resistance of urethane elastomers [2]. The relationship between polymer structure and the hydrolysis resistance of such materials, has been adequately dealt with in other publications [3].

The tests discussed in these papers involve the measurement of the physical properties of unsupported films and poromerics after various forms of accelerated hydrolysis. The author feels that some of the adverse criticism regarding the hydrolysis resistance of urethane coated fabrics might have been avoided if it had been made clear to the industry that the tests had been applied to relatively thick unsupported films.

When considering urethane coated fabrics, the question that must be asked is will the coated fabric be subjected to conditions of high humidity and elevated temperature. It is fair to say that most of the coated fabrics produced by the transfer process will not be subjected to such conditions. It is for this reason that many people in the industry believe that the problem of hydrolysis has been overemphasised. This holds true for furniture, garments and footwear where many faults that have been attributed to hydrolysis could equally be attributed to poor coating techniques, and the effects of exposure to light.

Some of the papers that have been published specifically on this subject of hydrolysis resistance of urethane coated fabrics have been misleading on the controversial issue of "one versus two" component tie-coats. There is no conclusive evidence to date to justify the general claim that on the grounds of molecular weight, one component tie-coats yield coated fabrics of a superior hydrolysis resistance to those fabrics comprising two component tie-coats. Given a suitable choice of raw materials, both one component and two component tie-coats can be produced that exhibit a wide range of hydrolysis resistance.

Light Stability

It must be appreciated that the "yellowing" of white and light coloured urethane coated fabric is not the only consideration when discussing 'light stability'. The loss of physical properties which usually accompany such discolouration must also be examined.

The stabilisation of a urethane elastomer can be achieved in two ways: by the incorporation of additives, such as U.V. absorbers and antioxidants, or by the synthesis of an inherently more stable elastomer.

The use of additives to impart improved light stability is still the most popular approach though in the last year or so the use of so called 'non-yellowing' urethanes has gained momentum. Reasons for the slow growth of the non-yellowing type of urethane elastomers include availability and the high price of the basic raw materials and doubts regarding the physical and chemical properties of these polymers.

We have subjected coated fabrics to three types of light exposure tests, namely:

(1) Window exposure (behind glass facing West)
(2) Xenotest exposure (30°C/65% RH)
(3) Light Box (*ca.* 40°C/45% RH)

The Light Box is a piece of apparatus which consists of a 500 Watt Mercury tungsten fluorescent lamp enclosed in a cylindrical box. The samples to be tested are placed on the inside wall at a distance of 20 cms away from the centre of the bulb. Giles and co-workers [4] have shown that the spectrum of the mercury tungsten fluorescent lamp is very close to that of the Xenon lamp.

After the light exposure tests the flex performance of the samples are assessed using either a Bally or Schildknecht flex tester.

The first set of results for discussion (see Table 6) are for a series of coated fabrics exposed in the Light Box, Xenotest and in the Window respectively. After analysis of the results and visual examination of the coated fabrics, the following conclusions can be drawn.

1. On average the flex performance was inversely proportional to the extent of yellowing. Bearing in mind the difficulties associated with the reproducibility of flex testing the following relationship between the severity of the exposure tests was established:

   Two weeks Light Box > One week Light Box ≃ two weeks Xenotest > ten months Window > one week Xenotest > two months Window.

2. For coatings not containing additives the degree of yellowing and the flex performance is very much dependent on the level of pigmentation.

3. After two months Window exposure there was negligible degradation in all the coatings containing additives. Degradation was extensive in the case of clear coatings and proportionally less for the pigmented coatings.

4. After ten months Window exposure all coatings showed signs of yellowing. Again the extent of yellowing in coatings containing pigments and additives was minimal whilst yellowing in those containing additives only was significant.

Thus from the results of the window exposure it can be deduced that the effective lifetime of the additives is of the order of twelve months in clear coatings. With the present knowledge obtained from the Xenotest and Light Box exposure tests the effective life of the additives in the presence of pigment is estimated to be of the order of 2–3 years for coatings exposed in direct sunlight behind glass (see Table 6). It must be appreciated that with the exception of upholstery applications in cars, coatings would not be subjected to such severe exposure in normal domestic use. Finally it is worth noting that all the coated fabrics performed well on the FIRA hydrolysis test thus showing once again that the relevance of accelerated hydrolysis testing for upholstery applications is questionable.

Because the effect of exposure in the Light Box and the Xenotest shows the former to produce greater degree of degradation at lower relative humidity but higher temperature, it was deemed prudent to study the effect of temperature alone on the degradation of the same series of urethane coated fabrics.

The heat stability results appear in Table 7 alongside the light stability results on the corresponding fabrics. Significant decrease in flex performance only occur after heating at 140°C for one week, although colour changes are more marked even at 120°C than those of light exposed samples. No significant cracking was however observed on flexing after exposure to heat at 120°C. From our work on light exposure we observed that colour changes was indicative of a marked change in physical properties but apparently this is not so in the case of heat exposure.

Development of Future Aging Tests

One of the main deficiencies that has been apparent since the introduction of urethane elastomers of the type used in the transfer process is their susceptibility to breakdown when exposed to natural or artificial weathering conditions. In natural weathering conditions the elastomers could be exposed to light, air and high humidity conditions.

It has always been recognised that it is the combination of these factors that contributes to the breakdown. In a diverse field such as the coating industry, where the end products are exposed to widely different conditions, then inevitably one factor is singled out as being the most important.

In an attempt to devise a suitable general aging test for urethane coated fabrics, samples of both aromatic and aliphatic based coatings were subjected to the following series of tests.

1. **Light Exposure Tests.**–including 40 weeks Window exposure; 1 week Xenotest exposure and 1 week Light Box exposure.

2. **Hydrolysis Tests.**–40 weeks immersion in tap water, distilled water, aqueous adipic acid solution (1.2 g/l) and aqueous urea solution (1.2 g/l) at room temperature.

Table 6. Effect of Light Exposure on P.U. Coated Fabrics

Ref. No.	Top Coat	Tie Coat	XENOTEST		LIGHT BOX		WINDOW	
			1 week	2 weeks	1week	2 weeks	8 weeks	40 weeks
HG 14	S1 (clear)	A1 (one - component)	2,000(G4)	-	3,000(G4)	40(G2)	2,000(G3)	400(G3)
HG 15		A2 (two - component)	4,000(G1)	-	4,000(G2)	800(G2)	9,000(G2)	1,200(G3)
HG 16	S1 + 5% White Pigment.	A1	10,000(G3)	-	2,000(G3)	4,000(G3)	100,000(G2)	300(G3)
HG 17		A2	23,000(G4)	5,000(G2)	35,000(G3)	23,000(G2)	164,000(G2)	6,000(G2)
HG 18	S1 + 10% White Pigment.	A1	16,000(G2)	10,000(G2)	25,000(G3)	5,000(G2)	27,000(G1)	2,000(G2)
HG 19		A2	30,000(G1)	28,000(G1)	28,000(G2)	30,000(G2)	200,000(G1)	19,000(G2)
HG 23	S1 (clear) + additives.	A1 + additives	100,000(G2)	8,000(G3)	27,000(G3)	4,000(G2)	200,000(G0)	20,000(G3)
HG 24		A2 + additives	200,000(G0)	11,000(G2)	-	10,000(G2)	200,000(G0)	53,000(G3)
HG 25	S1 + 5% White Pigment + additives.	A1 + additives	100,000(G1)	9,000(G2)	-	4,000(G2)	160,000(G0)	50,000(G2)
HG 26		A2 + additives	115,000(G2)	-	228,000(G0)	83,000(G2)	240,000(G0)	56,000(G2)
HG 27	S1 + 10% White Pigment + additives.	A1 + additives	32,000(G2)	12,000(G2)	60,000(G2)	24,000(G3)	110,000(G0)	25,000(G1)
HG 28		A2 + additives	100,000(G2)	100,000(G2)	145,000(G2)	100,000(G2)	200,000(G0)	200,000(G1)

NOTE: 1. All the samples were undamaged after 100,000 cycles before and after the FIRA hydrolysis test.

2. All samples were flexed on a Bally Flexometer.

Table 7. Comparison of Light and Heat Stability of P.U. Coated Fabrics

Ref No	Top Coat	Tie Coat	Initial	1 Week Light Box.	1 Week Oven Treatment. 60°C	120°C	130°C	140°C
HG 14	S1(clear)	A1 (one-component)	2.0 (G1)	0.010 (G3)	1.0 (G2)	1.65 (G0)	0.9 (G3)	0.200 (G2)
HG 15		A2 (two-component)	2.0 (G0)	0.039 (G3)	1.0 (G0)	1.65 (G0)	1.0 (G1)	1.0 (G2)
HG 16	S1 + 5% White Pigment.	A1	2.0 (G1)	0.038 (G3)	1.0 (G1)	1.65 (G1)	1.0 (G2)	0.200 (G2)
HG 17		A2	2.0 (G0)	0.170 (G2)	1.0 (G1)	1.65 (G1)	1.0 (G0)	1.0 (G1)
HG 18	S1 + 10% White Pigment.	A1	2.0 (G2)	0.060 (G2)	0.7 (G2)	1.65 (G1)	1.0 (G0)	0.500 (G2)
HG 19		A2	2.0 (G0)	0.300 (G1)	1.0 (G1)	1.65 (G0)	1.0 (G0)	0.200 (G2)
HG 23	S1 (clear) + additives.	A1 + additives	2.0 (G0)	0.172 (G3)	1.0 (G0)	1.65 (G0)	1.0 (G0)	1.0 (G0)
HG 24		A2 + additives	2.0 (G0)	0.200 (G2)	1.0 (G0)	1.65 (G0)	1.0 (G0)	1.0 (G1)
HG 25	S1 + 5% White Pigment + additives.	A1 + additives	2.0 (G0)	1.0 (G1)	1.0 (G1)	1.65 (G0)	1.0 (G0)	1.0 (G0)
HG 26		A2 + additives	2.0 (G0)	1.3 (G1)	1.0 (G0)	1.0 (G0)	1.0 (G0)	1.0 (G0)
HG 27	S1 + 10% White Pigment - additives	A2	2.0 (G4)	0.172 (G1)	1.0 (G2)	1.0 (G0)	1.0 (G0)	1.0 (G1)
HG 28		A2 + additives	2.0 (G0)	1.5 (G1)	1.0 (G0)	1.0 (G0)	1.0 (G1)	0.5 (G2)

NOTE: (1) All samples were flexed on a Schildknecht Flex Tester. Figures represent Number of cycles x 10^6

After the exposure trials the samples were flexed and the results are presented in Table 8. From these results, the following conclusions can be drawn.

a. The aliphatic based coatings are inferior to the aromatic based coatings in their resistance to hydrolysis. It is interesting to note that immersion of the coatings in adipic acid solution rather than in a solution of urea has resulted in most degradation. Hence it might be valuable that accelerated hydrolysis tests at elevated temperatures (if such a test is required) be conducted in an acidic solution e.g. adipic, as well as a solution of urea.

b. With the exception of aromatic systems that were pigmented black, the aliphatic systems have exhibited superior light stability after 40 weeks' window exposure tests. This was not unexpected since aliphatic based elastomers are inherently more stable. The excellent results obtained on the black aromatic based samples confirm the well established fact that carbon black acts as a good ultraviolet absorber.

c. Regarding the accelerated light aging tests, it can be seen that results from the Light Box correlate better with 40 weeks' window exposure than results from the Xenotest.

CONCLUSIONS

Consideration of the results obtained from application of the proposed general aging tests for urethane coated fabrics leads to the conclusion that the following accelerated tests could be significant in predicting their durability in various applications.

Upholstery

Light Exposure	Bally Flexometer	Schildknecht
NONE	150,000 cycles(minimum) no damage.	1 million cycles (minimum)no damage
Either 2 weeks Xenotest(30°C/65% RH) or 1 week Light Box (40°C/45% RH).	Minimum 70,000 cycles no damage.	Minimum 500,000 cycles

Footwear

Hydrolysis Conditions	Bally Flexometer	Schildknect or SATRA VAMP.
NONE	150,000 cycles(minimum), no damage.	1 million cycles minimum, no damage.
Tropical Test 2 weeks at 70°C, 95% relative humidity.	150,000 cycles, no damage.	1 million cycles, no damage.

Table 8. *Long Term Aging Tests on P.U. Coated Fabrics*

Ref. No.	Top Coat	FIRA Hydrolysis	Window Exposure 40 weeks	1 week Xenotest Exposure	1 week Light Box Exposure	Room Temperature Hydrolysis Test (40 weeks)			
						Aqueous Urea Solution (1.2 g/1)	Tap Water	Distilled Water	Aqucous adipic acid solution (1.2 g/1)
DW 1	TC1 (Off White)	GO	10,000(G2)	200,000(GO)	-	GO	GO	GO	GO
DW 2	TC2 (Light Tan)	GO	24,000(G2)	GO	-	GO	GO	GO	GO
DW 3	TC2 (Dark Brown)	GO	23,000(G3)	GO	-	GO	GO	GO	GO
DW 4	TC3 (Black)	GO	250,000(GO)	210,000(GO)	216,000(GO)	GO	GO	GO	GO
DW 5	TC3 (Black)	GO	250,000(GO)	320,000(GO)	350,000(G1)	GO	GO	GO	GO
DW 6	TC4 (Oyster)	GO	30,000(G2)	110,000(G1)	45,000(G2)	GO	GO	GO	GO
DW 7	TC5 (Oyster)	GO	G1	24,000(G1)	28,000(G2)	GO	GO	GO	GO
DW 8	TC6 (White) (Aliphatic)	GO	63,000(G1)	120,000(G1)	80,000(G1)	G1	50,000(G1)	GO	50,000(G2)
DW 9	TC6 (Oyster)	GO	250,000(GO)	143,000(G1)	GO	-	-	-	-
DW 10	TC6 (Oyster)	G2	50,000(G1)	120,000(G2)	120,000(G2)	GO	GO	G1	G3
DW 11	TC6 (Oyster)	G3	G1	67,000(G1)	80,000(G1)	G1	G4	GO	G3

NOTE: 1. The flex performance was assessed using a Bally Flexometer and unless otherwise stated the samples were flexed to 100,000 cycles.

2. All the samples were undamaged after 100,000 cycles before aging.

Garments

This is a less demanding application than either upholstery or footwear and the following tests are proposed.

Light Exposure	Schildknecht
NONE	1 million cycles, minimum, no damage.
1 week Xenotest or 3 days Light Box exposure.	500,000 cycles, minimum, no damage.

ACKNOWLEDGMENTS

The author wishes to thank all his colleagues for helpful discussions. Finally, he wishes to thank the Directors of Lankro Chemicals Limited for permission to present this paper.

REFERENCES

1. F. McGarry, Cabinet Maker, March 17, 1972.
2a. G. S. Schollenberger and F. D. Stewart, *Journal Elastoplastics*, Volume 3 (1971), p. 28.
2b. L. G. Hole and S. G. Abbott, "The Chemical Stability of Polyurethanes in Artificial Leather," Inter-Sat 1971 (SATRA SYMPOSIUM).
3. H. J. Koch, *Journal of Coated Fibrous Materials*, Volume 1 (1971), p. 118.
4. G. H. Giles and D. Baillie, Journal of the Society Dyers and Colourists, September 1969, p. 410.

HENRY R. LASMAN

Pandel Bradford, Inc.
200 Main St., Lowell Mass. 01852

VINYL AND URETHANE COATED FABRICS FOR SHOE UPPERS*

(Received December 27, 1974)

1. INTRODUCTION

VINYL AND URETHANE coated fabrics account for the predominant share of man-made shoe uppers. According to a recent study by The American Footwear Industry Association (AFIA), man-made materials for shoe uppers will be used in 50% of all shoes made in the United States this year. Assuming a domestic production in the order of 500 m pairs, almost 25 m linear yards or 350 m sq. feet of artificial leather will go into shoemaking; over 90% of this quantity will be taken by vinyl and urethane coated fabrics, and, if the past five years are any indication of the future, vinyl and urethane coated fabrics will continue to claim an ever-increasing share of the shoe upper market.

Today, vinyl and urethane coated fabrics are used in a wide variety of styles which range from casual women's shoes to men's work boots, from children's dress shoes to waterproof footwear, from house slippers to knee-high boots. These fabrics are made into shoes by several common shoe construction techniques, although they are preferred in Compo, slip and string lasted shoes. Artificial leathers, particularly vinyls, have given rise to unique labour-saving shoemaking operations. A prime example of these is the high-frequency flow-moulding of shoe uppers.

Vinyl and urethane coated fabrics are, therefore, no longer a novelty of passing importance to the shoe manufacturer. Rather they are firmly established shoe upper materials because of their aesthetic values, technological utility, and economic importance.

The objective of this paper is to acquaint the shoe manufacturer with the vinyl and urethane fabric technology and to offer a survey of the products generated by this technology.

*This paper was originally presented at SATRA – North America's Conference on Man-Made Shoe Upper Materials, Toronto, Canada, May, 1974. Permission to reprint was given by Shoe Materials Progress, August, 1974.

2. MANUFACTURING METHODS

The production of vinyl and urethane coated fabrics encompasses the following steps: compounding, coating, finishing and inspection, including testing.

Compounding

In the compounding step a homogeneous blend of all ingredients of the coating composition is produced.

In the case of vinyl compounds this involves the blending of vinyl resins with plasticisers, stabilisers, pigments and occasionally blowing agents, if a cellular coating is to be obtained. Depending on the coating method, the vinyl compound can be a liquid dispersion, called a plastisol, in which the PVC resin particles are suspended by agitation in the plasticiser. The conversion of a plastisol into a fused, elastomeric material occurs in the subsequent coating and heating operations. Alternately, a pre-blend consisting of vinyl resins, plasticisers and other ingredients can be blended in an intensive, high shear mixer, in which under the influence of heat, the resin blend is converted into a fused, elastomeric compound; this fused compound is then applied to the fabric backing.

Urethanes are usually applied to shoe upper fabrics from solutions containing a carefully selected blend of polymers, pigments, protective agents and other additives.

Before the application to a substrate, the coating compound is subjected to an in-process quality control which includes a colour comparison with the standard, a determination of basic processing characteristics, and a check for compliance with standard compound properties.

Coating

The application of the coating compound to a substrate, commonly a continuous textile web, is accomplished by one of the three basic techniques: casting, calendering and melt-extrusion. Of these, the casting method has played a dominant role in the manufacture of both vinyl and urethane shoe upper materials; also, the calendering method is extensively practised in the manufacture of vinyl uppers, whilst the extrusion process has a limited applicability in this field.

The manufacture of vinyl or urethane fabrics by the casting method entails the application of a liquid coating, either a plastisol or a urethane solution, in the form of a smooth, wet film of uniform thickness to a substrate. The coating can be applied directly to the fabric backing, or it can be cast on a paper carrier and subsequently transferred to the fabric by lamination.

The *direct coating* method is schemmatically depicted in Figure 1. Several different wet film applicators, so-called coating heads, are in use. All of these operate with coating knives which, depending upon the support under the substrate or transfer paper, are called knife-over-blanket, floating knife, or knife-over-roll

(Figure 2). The plastisol or solution coated web is carried through a hot air oven where the applied film is solidified as the plastisol fuses or the solvent evaporates. The coated fabric is cooled and wound up into a roll. To obtain a sufficient polymer deposition, this process is repeated several times, frequently in a tandem arrangement with several coating heads.

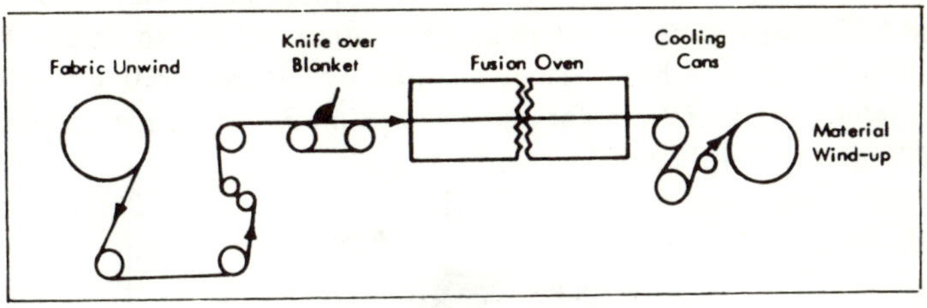

Figure 1. Direct coating unit.
(Source: Melliand Text. Rep 1963)

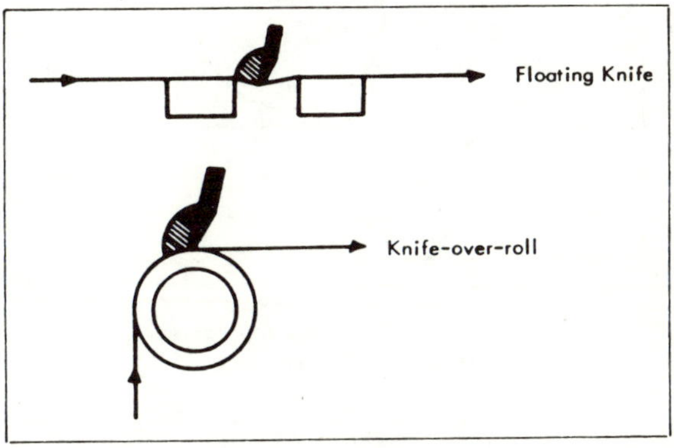

Figure 2. Coating knives.
(Source: Melliand Text. Rep 1963)

Usually the first coating is compounded and applied so as to fill, as much as possible, the interstices of the woven textile. This provides an impervious smooth surface for subsequent, more flexible coatings.

The final coating is an abrasion and light resistant wear layer which can be readily decorated in the finishing operation.

Because the textile backer is filled and sometimes even impregnated with the coating, the fabric mobility is decreased, resulting in a product that is stiff and boardy. Although the adhesion of the first coat – the tie coat – to the substrate is excellent, the multiple layer structure gives rise to potential delamination problems between layers, particularly when these, of necessity, are of different composition and flexibility. Today the use of the direct coating method is limited to the production of certain grades of vinyl patent upper materials. With the trend towards softer man-made uppers, the direct coating method, the oldest textile coating technique, has been superseded by a more versatile transfer casting method.

In the *transfer coating method* the plastisol or urethane solution is cast on a release paper carrier by means of knife-over-roll (Figure 3) or a reverse roll coater (Figure 4). The reverse roll coater which applies the coating by a wiping action affected in the opposite direction of the carrier movement, affords better gauge control and higher operating speeds than a knife-over-roll coating head. The latter performs better in casting heavy gauge coatings or fast evaporating solution coatings. A complete, modern train for the production of vinyl shoe upper fabrics is schematically depicted in Figure 5.

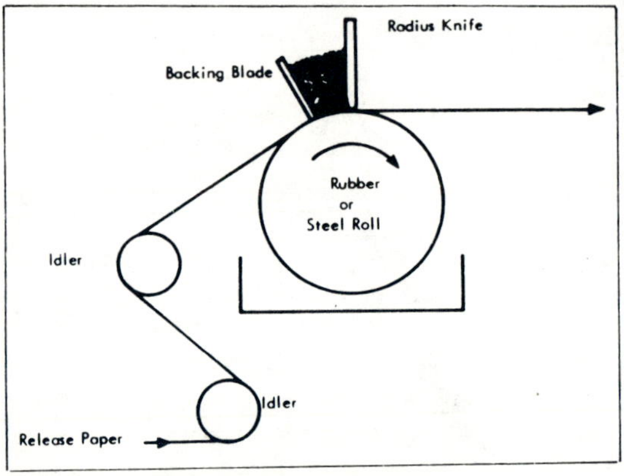

Figure 3. Knife-over-roll coater.
(Source: S D Warren Co.)

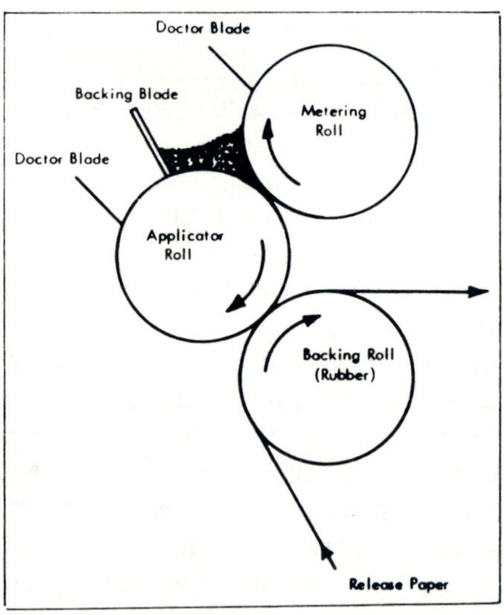

Figure 4. Three roll reverse roll coater.
(Source: Mod. Plastic Encyclopedia 1958)

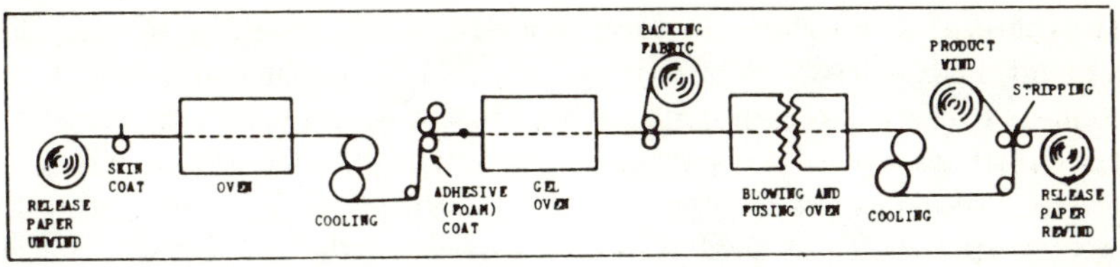

Figure 5. Casting-on-paper process for expanded PVC production.
(Source: S D Warren Co.)

After passing through the first coating head, the paper carrier with a wet layer of plastisol, the skin layer, enters the first oven where, under the influence of heat, the vinyl resin particles completely absorb the plasticiser, forming a tack-free but still heterogeneous, brittle film.

Upon cooling, a second layer, either dense or foamable, is applied to the gelled skin layer and, through moderate application of heat, the viscosity of the second layer is increased to the point that it becomes tacky. This layer acts as a hot melt adhesive for the fabric backing which is combined in a laminator at the oven exit. The laminate, still on the paper carrier, enters a long hot air oven where a complete fusion of the plastisol takes place. The fusion process produces a complete solvation of the resin and a melting of the layers into a homogeneous, strong elastomeric coating. If a foamable compound is used as the adhesive layer, a cellular structure is produced during the fusion process. After the fusion step, the vinyl fabric passes over a series of cooling cans which sufficiently cool the vinyl coating to permit easy release from the carrier. The vinyl fabric, wound in large master rolls, is then ready for the finishing operation.

The transfer casting of urethane fabrics follows essentially the same steps. Unlike plastisols, however, urethane solutions produce strong, elastomeric films immediately after the passage through the hot air ovens, where the solvent evaporates. The lamination to the fabric is carried out immediately upon the application of the adhesive layer.

In the transfer coating method the lamination of the fabric occurs while the coating is still on the carrier and, therefore, the adhesive penetration into the fabric can be carefully controlled. Indeed, only through this lamination technique it is possible to obtain highly flexible, soft shoe upper materials having unique aesthetic properties and characteristic leather-like drape.

The proper operation of the many steps involved in the transfer-casting method calls for numerous controlling devices which measure and record the gauge, the line speed and the temperatures of the individual ovens or zones within the ovens. The quality of the continuously produced coated fabric is monitored by a rigorous in-line testing of representative samples from individual master rolls to determine the adhesion of the coating to the substrate, the integrity of the coating, and the physical and mechanical properties of the product.

The transfer coating method affords great production versatility since it is adaptable to highly plasticised vinyl coatings, to vinyl dispersions with low plasticiser content, and to urethane solutions; fabrics ranging from knits to sateens and non-woven substrates can be conveniently coated. The transfer process allows quick change of colours, gauges and types of fabric construction. These processing advantages combined with the highly desirable quality of the end product, make the transfer coating method the most widely used coating technique for producing man-made shoe upper materials.

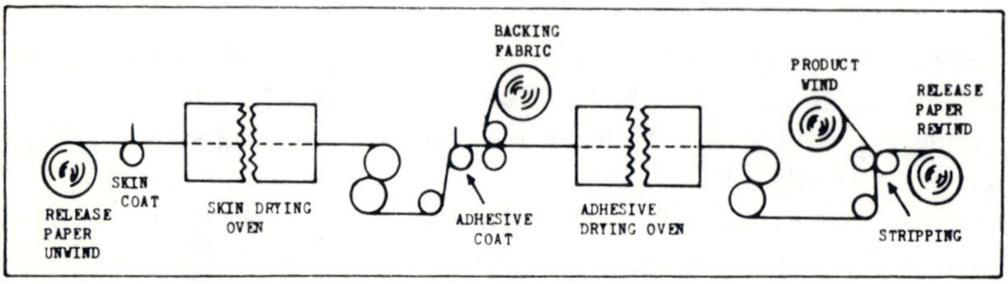

Figure 5A. Casting-on-paper process for urethane production.
(Source: S D Warren Co.)

The *calendar coating* operation involves the application of an unsupported calendered sheeting to a textile substrate by means of heat and pressure. The diagram of the calendering operation is depicted in Figure 6. The process of calendering converts the fluxed PVC compound into a continuous sheet of controlled thickness and smooth surface. This operation is performed between two or more pairs of rotating, heated cylinders, called calender rolls. Each roll pair forms a rotating die through which the calendered material is drawn. Several types of roll arrangements are available: the most popular are the inverted "L" and the inclined "Z" calenders (Figure 7).

In particular, the latter provides a shorter path for material travel and permits easy accessibility for coating fabrics. The calendered sheeting is combined with the backing by forcing the fused compound into the interstices of the fabric by means of controlled heat and pressure in a nip formed by two rotating rolls (Figure 8).

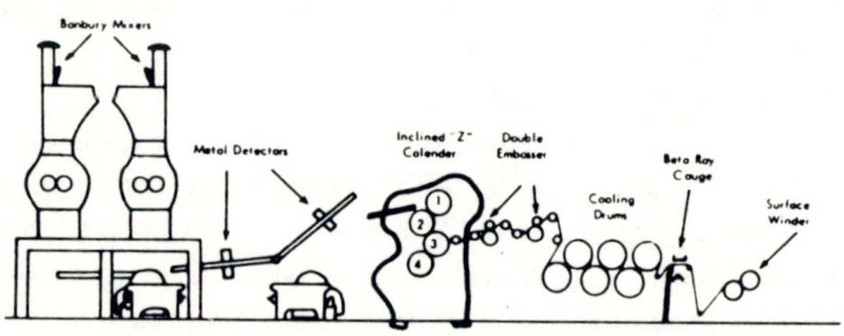

Figure 6. Calendering train for production of vinyl fabrics.
(Source: Mod. Plastic Encyclopedia 1958)

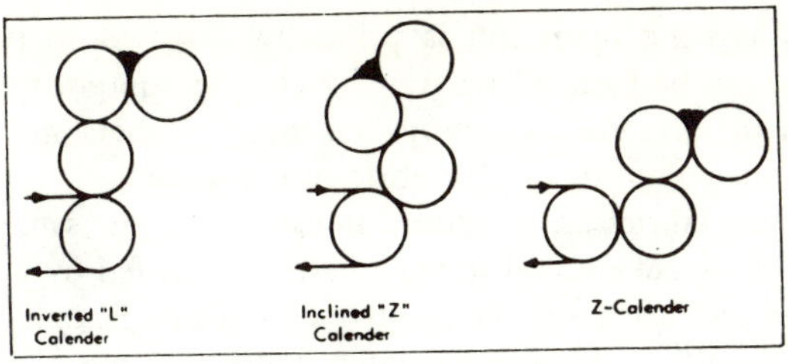

Figure 7. Calender configurations.
(Source: S D Warren Co.)

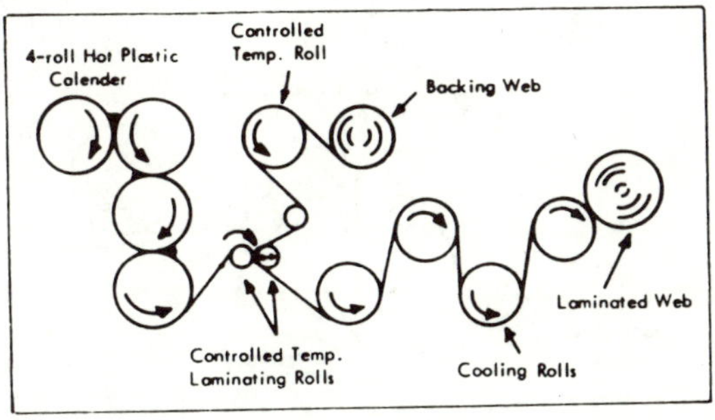

Figure 8. Coating by 4-roll system of post calender lamination.
(Source: Mod. Plastic Encyclopedia 1958)

Alternately, calendered webs can be laminated to fabrics by means of an adhesive as shown in Figure 9. When expanded vinyl fabrics are produced, a foamable sheeting is laminated to the fabric and the laminate is passed through an expansion oven where the heat triggers off the decomposition of the blowing agent previously incorporated in the coating compound. The supported cellular layer is laminated to a dense skin either before or after the expansion step.

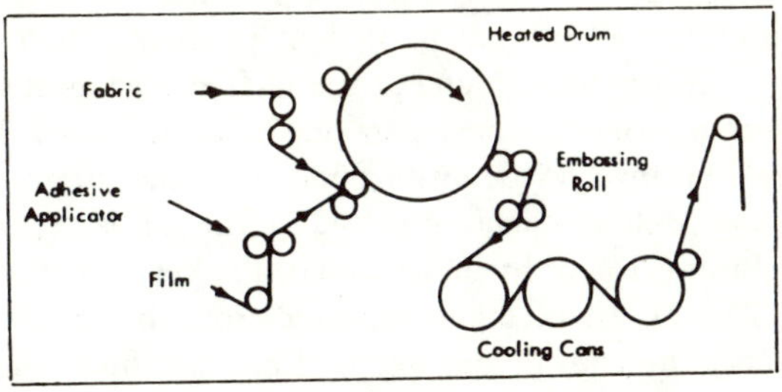

Figure 9. Lamination of vinyl film or sheeting to fabric backing.
(Source: Wellington Sears Hand. of Ind. Text. 1963)

The calender coating operation is primarily designed to process stiffer vinyl compounds than can be handled on transfer coating equipment; since the product changeovers and the attendant clean-ups are more cumbersome, it is most suitable for long runs of one colour in a given fabric construction.

Extruded coated fabrics are produced in essentially the same way as calendered laminates. Instead of an internal mixer, the extruder, fed by a vinyl dry-blend, is fluxing the blend and discharges it through a flat sheet die in a form of continuous vinyl web. The hot film is then immediately fed into a nip of a two roll laminator where it is combined with the substrate (Figure 10).

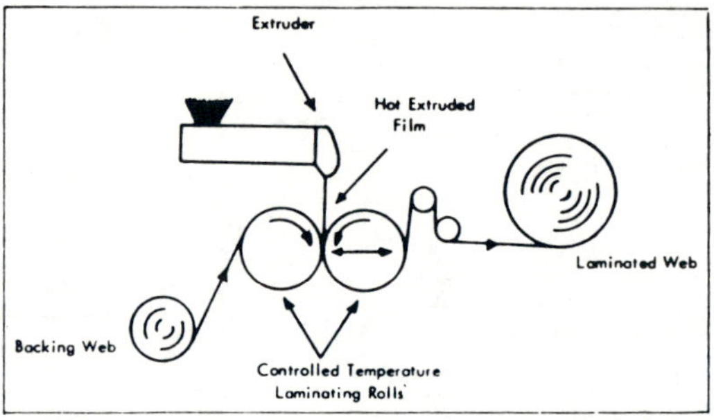

Figure 10. Coating of fabrics by extrusion coating.
(Source: Mod. Plastic Encyclopedia 1958)

Finishing

More than any other single factor, surface decoration of vinyl and urethane coated fabrics has created a permanent market for these products in the shoe industry.

Decorating methods, like printing, spanishing (see later), top-coating and embossing convert a flat, dull looking piece of coated fabric into a richly grained, three-dimensional, eye-appealing man-made shoe upper material having the look, the drape and the hand of the natural leather product.

Vinyl and urethane fabrics are printed to produce colour effects and designs against the solid, uniform background of the surface. The most widely used printing method is gravure printing, although spraying is also practised. The gravure printing method (Figure 11), involves depositing ink onto the surface to be printed from cell-like crevices of a photo-engraved cylinder. The ink is deposited in the cells by contact with a furnish roll or by immersion of the lower portion of the print roll in an ink trough. The excess of ink is removed from the roll by means of a doctor blade, so that only the cups of the engraved area are filled with ink. The printing roll is then brought into contact with the vinyl or urethane surface and the ink is transferred, under moderate pressure, from the engraving to the coated fabric.

Gravure cylinders etched with an entire screen pattern are used to uniformly

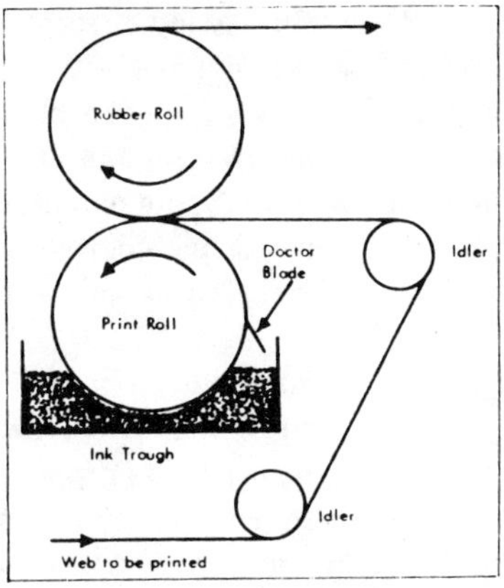

Figure 11. Direct gravure printer.
(Source: Mod. Plastic Encyclopedia 1958)

apply a variety of top-coatings over the entire surface area of the sheeting. These top-coatings may be clear or translucent, bright or dull, with a high degree of slip or greasy feel, depending upon the requirements dictated by stylists and customers.

Random aniline pattern can be applied to vinyl and urethane surfaces, just as to tanned leather surfaces, by means of spray guns, either reciprocating or rotating.

Another method borrowed from the leather industry has found application in producing very fine grained effects on man-made shoe upper materials. Called "spanishing" the method entails wiping of the ink into the crevices of embossed sheeting. Spanishing produces fine, leather-like two-tone effects, unobtainable by any other decorating method.

Leather-like grains are most frequently imparted to flat man-made upper materials by embossing. In this operation, a preheated thermoplastic sheeting, e.g., vinyl coated fabric, is continuously fed into a nip formed by a rubber back-up roll and an engraved, internally cooled, embossing roll. Under the influence of heat and pressure, the surface of the coated fabric takes on the reverse image of the engraved roll, i.e., peaks of the engraving correspond to valleys in the final product. The grain is set through a rapid cooling of the material on the embossing roll. Embossing of coated fabrics can be carried out as a part of a multiple step finishing operation, or it can be set-up in line with the fabric coating operation, prior to printing or spraying. The embossing method is applicable to all thermoplastic coated fabrics, whether produced by casting or calendering; it is capable of reproducing minute grain details with a high degree of fidelity.

Alternatively, embossed vinyl and urethane fabrics can be obtained by transfer casting on pre-embossed release papers. This method is particularly useful in the manufacture of soft materials which would be easily crushed in the nip of an embosser.

Still another method of developing grained surfaces relies on a technique adapted from the leather technology known as "milling" or tumbling. A flat man-made web of a special construction can be continuously tumbled so as to break the smooth surface into a multitude of random veins and creases.

Conversely, a mirror bright surface simulating patent leather can be produced by embossing with a highly polished chrome plated roll, or by embossing through a polished film of Mylar or polypropylene. Casting on special release papers with high gloss surface leads to the same effect.

These are some of the decorating methods common to the coated fabrics industry and applicable to man-made uppers. Through combination and permutation of the various steps, through variation of print and embossing patterns, and through the development of new colours, the shoe manufacturer is assured of a never-ending stream of new and exciting shoe upper materials.

In the finishing operation, the quality assurance function encompasses checks on the grain retention and on the adhesion of printing ink and top-coating, as well as the visual and instrumental comparison of colour and gloss with the standard.

Inspection and Testing

From the finishing department, properly identified and documented master rolls are transferred to the inspection department. There the finished product is examined on inspection frames under standardised lighting conditions. As a given material passes the inspection table, its appearance is checked against the standard established at the time of the product's inception. Defective material is being cut out as master rolls are being broken into smaller shipping units. Random samples are submitted to the Quality Assurance Laboratory for final testing of physical properties such as Bally Flex resistance, low temperature impact resistance, abrasion resistance, tensile and tear properties.

Records are being made as to the disposition of the inspected material and are filed together with retained samples and quality assurance data for future reference.

3. PROPERTIES OF COATINGS

Vinyl chloride and urethane polymers are particularly suited as shoe upper materials. Coatings based on these polymers combine toughness with flexibility and exhibit an excellent balance of physical properties including high abrasion resistance; they feature ease of processing and lend themselves to a wide range of decorative techniques. Both vinyls and urethane coatings can be prepared in an unlimited pallet of colours which will remain bright and fast throughout the useful life of the end product. In particular, vinyls can be formulated to produce unsurpassed white and pastel colours having excellent light and heat stability. Urethane coatings are still somewhat deficient, although lately great improvements have been made in this respect. Vinyl coatings are also superior to urethanes with regard to chemical resistance in that the former are virtually unaffected by weak acids or bases, while the latter are susceptible to hydrolysis at elevated temperatures over a period of time. Significant progress has been made recently to develop urethane polymers and additives to increase the environmental stability of urethane coatings used for shoe upper materials. Comparing dense polymers, urethanes possess a significant degree of moisture vapour transmission while vinyls are completely impervious to water vapour.

Another basic difference between these two most common shoe upper polymers is worth noting. Common to all vinyl coating compounds is the presence of plasticisers which upon fusion convert the hard, brittle vinyl resin into a flexible material of any desired degree of hardness. In fact, through a judicious choice of plasticisers, vinyl compounds can be made to remain flexible at extremely low temperatures and to exhibit superior flow properties. The fusion of a plastisol is a physical phenomenon; although tightly bound to the surface of resin molecules, the plasticiser does not enter into the molecular structure of the polyvinyl chloride resin, rather it acts as a molecular lubricant having a certain degree of internal mobility. This mobility increases with rising temperatures, and under the influence of light or in contact with other polymers or solvents. Therefore, a careful selection of the most compatible and permanent plasticiser is mandatory when vinyl compounds are designed for shoe uppers.

Unlike vinyl resins, urethanes are internally plasticised, i.e., the desired degree of flexibility and low temperature properties can be built into the molecular backbone, and plasticising additives are unnecessary. Through a similar molecular rearrangement the polymer chemist is capable of producing either thermoplastic compounds or crosslinkable urethanes that will not flow under the influence of heat. Typical properties of vinyl coating compounds are presented in Table 1 and those of urethane coatings are listed in Table 2.

4. PROPERTIES OF VINYL AND URETHANE UPPER SHOE MATERIALS

Vinyl and urethane shoe upper materials are three ply laminates consisting of a skin, an adhesive, and a textile substrate.

The skin, either continuous or, in the instance of suede, discontinuous, provides the decorative outer surface which takes on a leather-like appearance through the previously described finishing processes. More importantly, the skin also determines the utilitarian properties of the laminate, among others the flex and scuff resistance as well as the low temperature behaviour.

The adhesive or tie coat anchors the skin to the textile substrate. This layer can be either dense or expanded; in the latter case it not only bonds the skin to the substrate, but also adds bulk and firmness to the product.

The textile substrate imparts the mechanical strength necessary for many shoemaking operations, e.g., lasting; beyond that it affords the structural toughness and

Table 1. Physical Properties of PVC Skin Compounds*

Construction	Skin on 0.036" (0.9mm) GA	Skin on 0.065" (1.7mm) expanded
100 % Modulus, psi	1,150	1,350
Ultimate Tensile, psi	2, 700	2,800
Elongation at break, %	400	375
Hardness, Shore A	72	80
Low Temperature Impact Passes at °F	-35°F (-38°C)	-30°F (-35°C)

* Obtained on unsupported film, 10 mils thick

Table 2. Physical Properties of Typical Urethane Skin Compounds*

Type of construction	Bright patent	Soft, glovy
100 % Modulus, psi	1,100	610
Ultimate Tensile, psi	6,000	5,200
Elongation at break, %	700	400
Low Temperature Impact Passes at °F	-30°F (-35°C)	-35°F (-38°C)

* Obtained on unsupported film, 1.5-2.0 mils thick

resistance to deformation so important to the shoe wearer. Until recently, a 100% cotton sateen has been used almost exclusively as a substrate for vinyl shoe upper materials. The high thread count of this textile offers a relatively smooth coating surface, and good mechanical properties that are reasonably balanced in both directions. In the case of urethane fabrics, in addition to mechanical properties, the substrate provides a uniform dense nap which greatly contributes to the leather-like break and the surface appearance expected of the end product. The most common fabric for this application is a 100% cotton filling faced sateen in which the filling has been repeatedly napped and sheared to form a velvet-like surface.

The mechanical properties of vinyl or urethane coated sateen parallel those of the substrate itself. Coated fabrics display some increase in tensile and burst strength, but also a decrease of tear resistance. Tear resistance properties are always impaired when woven yarns are locked into a rigid matrix through impregnation with polymeric binders: the higher the degree of impregnation, the greater the loss of tear resistance.

Directly coated fabrics or calender laminated fabrics which employ a tie coat show more pronounced deterioration of tear resistance than transfer coated substrates.

Significant improvement of all mechanical properties accompanied by a steep increase of moisture absorption is attained by combining the vinyl or urethane coated fabric to a drill or sheeting reinforced lining (Tables 3 and 4). Vinyl and

Table 3. Physical Properties of Typical Vinyl Upper Material

	100 % Cotton Sateen Substrate	Soft Vinyl Upper Material		
		Uncombined 0.036" ga	Combined (1)	Combined (2)
Grab Tensile (lbs/inch)	110 x 80	115 x 85	216 x 135	271 x 190
Tongue Tear (lbs)	8 x 8	7 x 7	7 x 9	9 x 10
Burst Strength	200	230	420	520

Combined (1) : Urethane foam on sheeting, nylon tricot
Combined (2) : Drill and Rayon Faille

Table 4. Physical Properties of Typical Urethane Upper Material

	100 % Cotton Napped/Sheared	Urethane Coated Upper	
		Uncombined	Combined (1)
Grab Tensile (lbs/inch)	65 x 80	75 x 90	153 x 180
Tongue Tear (lbs)	8 x 7	7 x 6	8 x 12
Burst Strength (psi)	180	195	360

Combined (1) : Urethane foam on sheeting, Nylon Tricot

urethane shoe upper materials of this construction offer the most advantageous set of properties to both the shoe manufacturer and the consumer.

The need for a more economical product than the multiple layer shoe upper construction brought about a wide acceptance of expanded vinyls for unlined string-lasted shoes, and more recently in flow-moulding applications as well. To ensure proper shape-holding ability of the unlined shoe, the expanded vinyl material (0.065") must of necessity be more boardy than the regular (0.036") gauge product. The mechanical properties of the unlined upper material are substantially lower than those of the lined multi-layered construction (Table 5). The advent of polyester/cotton blends has changed the situation. A light (2.03 yds/lb) polyester/cotton sheeting affords mechanical reinforcement superior to that of a heavy (1.21 yds/lb) 100% cotton sateen, however, the physical properties of expanded vinyl on 2.03 polyester/cotton sheeting still fall short of those displayed by the drill reinforced combined vinyl upper. Currently the best mechanical properties in unlined vinyl shoe upper materials are obtained on 1.56 polyester/cotton sateen, as shown in Table 5.

Expanded vinyl uppers based on this new synthetic fibre substrate come very close to the more expensive combined construction with regard to grab-tensile and burst strength. More important, the unlined construction surpasses the lined upper in tear resistance properties. This significant improvement of tear resistance will obviate the need for reinforcement of tear prone sections of shoe uppers with multiple cut-outs or edge under-cuts.

Table 5. Physical Properties of Typical Expanded Vinyl Upper Materials (Gauge 0.065" (1.7 mm)

	100 % Cotton Sateen	75/25 Polyester Cotton Substrate		
		2.03 Sheeting	1.68 Drill	1.56 Sateen
Grab Tensile (lbs/inch)	125 x 100	170 x 160	200 x 200	250 x 220
Tongue Tear (lbs)	7 x 7	8 x 8	12 x 12	15 x 15
Burst Strength (psi)	280	340	405	475

In offering the improved, synthetic fibre reinforced shoe upper material, attempts have been made to preserve all the desirable properties of the all-cotton backer: however, some differences are unavoidable. With reference to lasting properties, the stress-strain characteristics of the shoe upper materials are of significance and these are compared in Figure 12. Polyester fabrics display a lower modulus, i.e., under a moderate stress they will elongate more than all cotton sateen. As the stress increases, they will continue to elongate under loads that would cause the all-cotton sateen to break.

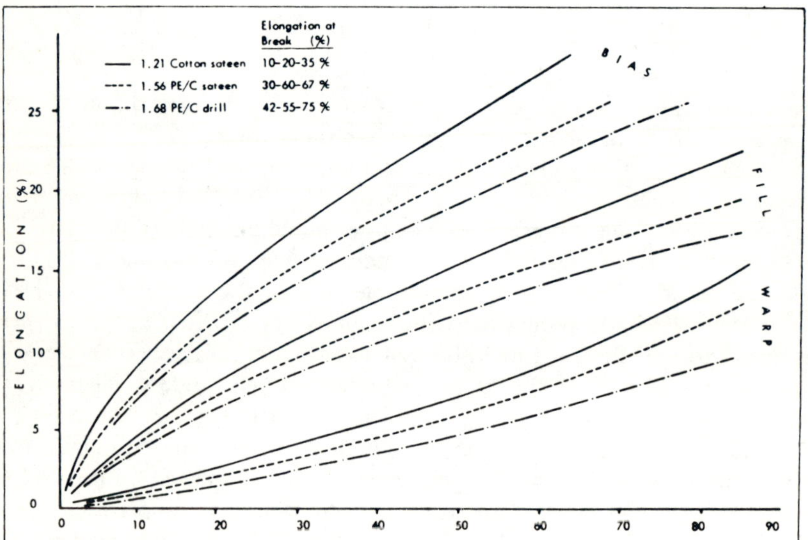

Figure 12. Elongation of Panda® 65

Other differences of importance to the shoe manufacturer are due to the hydrophobic nature of polyester fibres as compared to the hydrophilic properties of cotton. Inevitably, the moisture absorption of polyester/cotton blends is lower than that of the all-cotton fabrics, also, due to the differences in dye receptivity, in the direct or sulphur dye processes, only the cotton fibres are dyed whilst polyester fibres are stained off-white. Therefore, uniform colours, particularly black and dark brown, are not available at the present time. Several promising approaches are being worked on to obtain economical, crock-proof uniformly dark colours in polyester/cotton fabrics for shoe uppers.

Edge fraying, a problem common to all fabric backers can be overcome by a high degree of fabric saturation by the coating polymer but this impairs tear resistance properties. More successfully, edge fraying can be eliminated by the use of non-woven substrates. In particular, binder saturated, needle-punched non-wovens have drawn the attention of manufacturers of man-made uppers. The use of non-wovens in poromerics and linings is well established, but the application of non-wovens in vinyl coated shoe upper materials have met with little commercial success, probably because of high cost. In addition to fray resistance, expanded vinyls on non-woven substrates display leather-like lined appearance with good moisture absorption characteristics and excellent physcial properties as shown in Table 6.

Table 6. Physical Properties of Typical Vinyl Upper Materials on Non-Woven Substrates

Substrate	A	A	B
Ga. of Substrate	0.048" (1.2mm)	0.040" (1.0mm)	0.030" (0.8mm)
Total Gauge	0.060" (1.5mm)	0.060" (1.5mm)	0.070" (1.8mm)
Tensile Strength (lbs/inch)	235 x 237	185 x 215	95 x 110
Tear Resistance (lbs)	11 x 12	10 x 11	7 x 8
Burst Strength (psi)	325	320	180

The thermoplasticity of vinyls combined with a high power loss (dissipation) factor in a high frequency field resulting in heat generation within the material has led to the development of new production methods for shoe uppers. The power loss factor of plasticised PVC goes through a maximum between 5,000 and 50.000 herz. Electromagnetic effects in this frequency range causes an acceleration of molecular motion in the PVC compound with the attendant increase of internal temperature. Other materials of construction like aluminum, used for mould covers and dies, or silicone polymers, from which moulds are made, when placed in the same high frequency field, do not heat up because their dissipation factors are lower.

These differences in electrical behaviour are utilised in high frequency heat sealing and flow moulding. By applying unsupported vinyl sheeting to vinyl uppers through the high frequency heat sealing method, multicolour and fancy inlay effects are produced with unskilled labour and at high throughput rates. The shape of the inlay is defined by the cold metal die which seals the two hot vinyl materials around the periphery. To obtain strong, permanent weld lines both surfaces must be clean and free from slip agents and lubricants like silicones or stearates.

The flow moulding of vinyl upper materials is a process in which a piece of vinyl coated fabric is heated in a high frequency field and forced under low pressure against the surface of a silicone mould. Power requirements for flow moulding are much higher than for heat sealing since a much larger mass of PVC coating must be heated. In this process, a flat, pre-cut PVC upper can be completely textured, and stitching as well as perforations faithfully reproduced to match the original master that served to produce the silicone mould. Given a sufficient power source, any vinyl coated upper can be flow moulded provided that the mould depth at any given point does not exceed the thickness of the flat piece. When heavily textured uppers with pronounced welts are to be made, expanded vinyl fabrics of special construction are available. The thickness of such a construction is usually a fraction of the maximum mould depth; but when heat is applied, the internal pressure of the gas bubbles encapsulated in the foam layer causes the vinyl compound to completely fill the mould cavity and to faithfully reproduce the mould surface.

OUTLOOK

In spite of the great variety of vinyl and urethane fabrics currently available for shoe upper manufacture, the search for better and more economical leather substitutes continues unabated. Present efforts are aimed at the development of vinyl and urethane shoe upper materials with improved lasting properties, superior wear comfort, increased scuff resistance and generally better field performance. Economic considerations will dictate a more prudent utilisation of raw materials and this in turn will bring about the introduction of lighter yet stronger and more durable leather substitutes. The shoe manufacturer can look forward to a constant flow of vinyl and urethane fabrics that will provide him with economical alternatives to leather and will permit him a high degree of automation. With a rising standard of living throughout the world, the demand for products that traditionally have been made of leather, is fast outstripping the limited supply of hides. Increasingly, this gap can and will be filled by leather substitutes, foremost among these, vinyl and urethane coated fabrics.

ACKNOWLEDGMENT

The author is grateful to the management of Pandel/Bradford, Inc. for the encouragement and permission to publish this paper. I wish to express my appreciation to Tom Paine and the SD Warren Co. of Westbrook, Me. for the preparation of most of the slides and figures accompanying this presentation. Finally, the assistance of the Technical Department personnel of Pandel/Bradford, Inc. who have developed the data presented in the paper is thankfully acknowledged.

W. Kallert

Bayer AG
Leverkusen, West Germany

URETHANE FOAM FOR CARPET BACKING*

(Received December 27, 1974)

INTRODUCTION

Anyone who has witnessed the development and the resulting expansion in the applications of the urethanes is not particularly surprised by the appearance of urethanes in yet another new sector, in this case the carpet industry. This naturally gives rise to several questions of a general nature, the answer to which is practically an analysis of the situation in the carpet industry.

1. What is the present state of technical development in the industry into which urethane is to be introduced?
2. How do the established coating materials compare with urethanes?
3. Are there any additional aspects, such as environmental pollution, which influence the choice between the various coating systems?
4. How is the urethane to be formulated, to be produced and to be applied?

THE BACKING MATERIALS

Because the carpet industry in the whole world is so extraordinarily diverse, we will discuss mainly the situation in Western Europe. A carpet is a textile floor covering and we shall confine ourselves to discuss only tufted and needlepunched types in the form of tiles or as wall to wall carpet. At present the market share of tufted carpets is about 60%, needlepunched carpets make up about 20% of the market and the rest is mainly woven carpet which is usually not backed. The trend is towards more tufted floor coverings. About 85% of all tufted and needlepunched floor coverings are laid as wall to wall carpet, the rest as tiles. It was hoped that tiles would increase their share of the market but they are becoming gradually less important.

Figure 1 shows the cross section of a tufted carpet and demonstrates the various methods of anchoring and backing the underside. The diagram in the top right hand corner shows the absolute minimum treatment with an anchor coat. The third picture on the right shows the dominant coating treatment at present. The anchor

*Permission granted by Polymers Paint and Colour Journal, Polyurethanes Symposium – University of Surrey, September 23–25, 1974

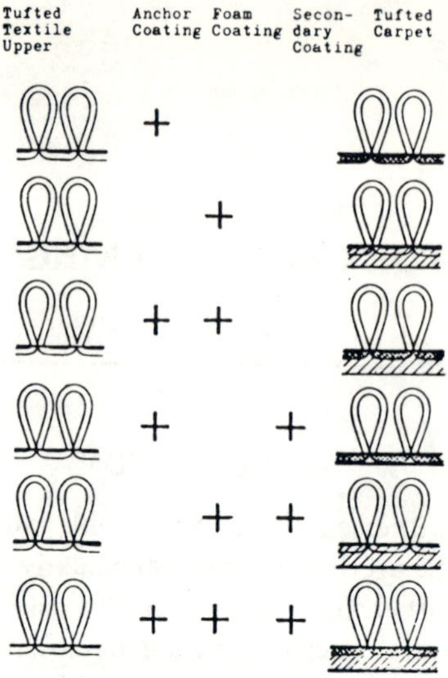

Figure 1. Various possibilities for backing tufted carpets with Desmoflex 2000.

coat transforms the textile upper into a usable carpet by anchoring the tufts into the tufting substrate. An additional foam backing prolongs the wear life and gives the carpet a luxury value. Foam is used preferably for wall to wall carpet and tiles are usually backed with a non-foamed material.

In the various European countries latex accounts for between 80 and 90% of all coating material for tufted and needlepunched wall to wall carpet. Small quantities of PVC and urethane are also used. Tiles are mostly backed with APP and other hot melts and, to a lesser extent, urethanes and PVC.

The present day plant for latex coating can accommodate rolls between 4 and 5 m wide and the coating is applied at a speed of 3 to 6 m/min. There is a very large new latex coating plant in Roermond in Holland which works with the exceptionally high throughput of 10 m/min. The processing with latex is effectively without technical problems and should continue to be the dominating process in the near future. The application of hot melts, for instance APP, allows a throughput of 10 to 15 m/min. This method is only suitable for solid backings and is used for the manufacture of tiles with a maximum width of 2 m.

Precise predictions about the long term technical trends in the production of textile floor coverings are not possible because new methods of producing the textile coating are known which have already achieved considerable importance in other continents and which will probably be gradually introduced into Europe. For these processes it might be necessary to develop completely new anchoring systems. This will require a coating system which is flexible in its application and properties.

PROCESS VARIABLES

Latex foam can be produced in a variety of qualities with differing density, elastic behaviour and tensile strength. Latex is applied by direct coating. During this process large quantities of water must be evaporated. This process naturally needs a relatively high amount of energy and to increase the throughput is difficult. The anchor coat, a highly-filled unfoamed formulation, is first applied. Shortly afterwards the so-called secondary backing, a low-filled mechanical foam, is put on.

Hot melts (preferable APP) are used almost only for the coating of tiles and cannot be foamed. They are applied on to a latex anchor coated carpet.

It is well known that urethanes are very flexible from the point of view of starting materials and also in their final properties. The systems that we offer contain no volatile components, have a very low energy consumption, require a low processing temperature and the plant occupies little space. Because urethane is applied by reverse coating, the thickness of the coating can be controlled within fine tolerances. It is possible to coat without an anchor coating for many applications. The technology causes no environmental pollution; the coating is odourless. The first full production plant has a working width of 5 m. This shows the great variability which the urethane chemistry has always promised for applications in this sector.

It may be seen that practically all the points that have been previously mentioned about the various systems interest the carpet backer and not the consumer. The processor is also responsible for solving the various environmental problems when using the various coating systems. Examples in other industries show that the expected energy consumption can play a decisive role in the choice between various technologies.

Other factors also to be considered include future fashion trends in carpets, new more economical carpet making machinery, and possible new government legislation in favour of the end-user. The backing material which is most suitable for fulfilling these many parameters will be preferred.

We need a urethane foam-backing with properties at least good enough for the required application. We can take the normal commercially available latex backing as a standard for comparison. The price per unit area must not be allowed to exceed the costs of comparable systems. The total production price is naturally the basis for the comparison. The possibility, for example, of producing a foam-backing without an anchor coat that has a satisfactory tuft anchorage for many applications, must also be taken into the price calculation.

A series of different recommendations based on urethanes has been published in order to answer this question.

One suggestion was to simulate the latex coating production and to use silicone stabilized mechanical foams which are applied with a knife, either in the direct, or in the reverse coating technique. Solidification is in the usual manner for urethane systems by application of heat helped by catalysts. The foaming agent is simply trapped air.

It has also been suggested that a reacting coating-mixture should be applied by spraying; that is in the direct coating technique. Foaming can be achieved with the usual water-isocyanate reaction. In order to obtain a uniform coating which is completely smooth it is necessary to have a relatively sophisticated plant.

Another possibility is to pour an activated reacting mixture to and fro on to a moving release belt which is smoothed off by guiding the carpet onto it and then solidified by means of heat. This method also uses the usual CO_2 development for foaming. With such a highly reactive mixture it is not possible to use a doctor blade. A precise adjustment is not possible particularly for very thin coatings.

Production Plant for Urethane Coating

Figure 2 is a schematic of the plant. Polyol and chalk are lead from storage tanks to a container to be mixed. The filled polyol is then mixed with isocyanate in a mixing head. A third component is added to achieve the desired density and physical properties. A suspension of pigment can also be added when necessary. One of the most distinctive features of this system is that the compound is applied directly onto the surface of a heated epoxy resin drum by means of a doctor blade. The distance between the doctor blade and the surface of the drum can be very precisely controlled. The entire drum rotates in an insulated housing which ensures a constant temperature. The carpet is guided tangentially onto the reactive, still liquid mixture. The movement of carpet and epoxy resin drum is synchronous. At the moment a throughput of 5 to 6 m of carpet/min is usual; the heavy backing for tiles with a speed of 3,5 to 4,5 m/min. Rolls of up to 5 m width can be accommodated; it is also possible to coat several narrower rolls at the same time. The temperature of the textile upper remains low because the temperature of the drum and its housing does not exceed 90°C. The carpet is then lead over a scray to be inspected, rolled up or cut into tiles and packed.

The Chemical System

The system is based on a specially formulated polypropylene ether and a liquid MDI. Polyester is not suitable because it cannot be expected to give the necessary service life in all conceivable cases. MDI was chosen because of its significantly lower vapour pressure in comparison to TDI during processing. The use of this isocyanate is without problems because industry has so much experience in handling it. The nature of the special catalyst allows a long pot life at room temperature, and the coating solidifies completely in about 90 sec at 80 to 90°C. The same basic formulation can be used to produce a non-foamed coating (density between 1.4 and 1.7 Mg/m^3 with 40% to 60% chalk), a lightly foamed coating (density approx. 1.1 Mg/m^3) and a low density foam coating (density approx. 0.35 Mg/m^3). Solid or only slightly foamed urethane coatings are compounded with a drying agent, e.g., zeolite. This prevents any formation of gas due to moisture in either the

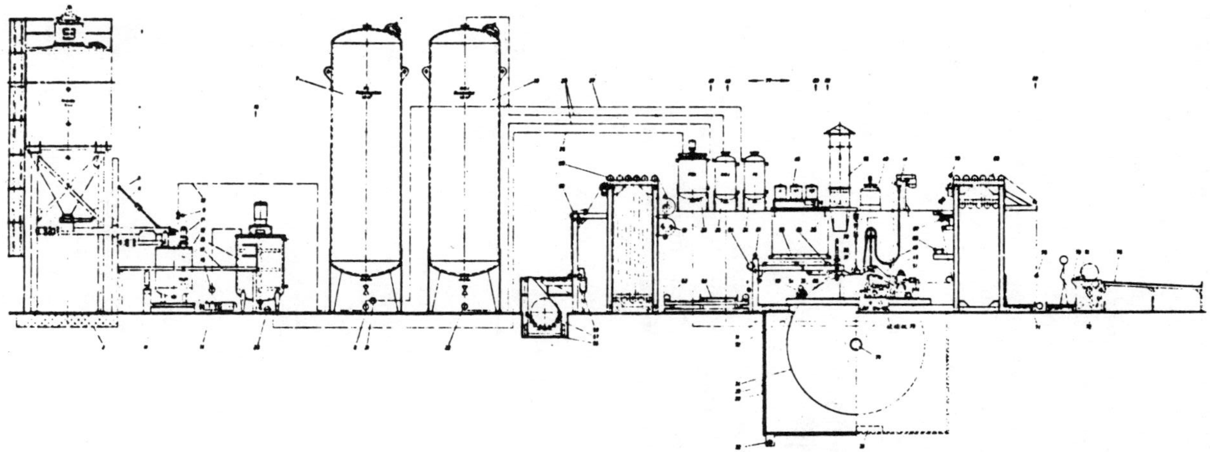

Figure 2. Complete PUR carpet backing plant (METZELER SCHAUM).

Key to Figure 2
(For simplicity only the more important parts are named in English)

72	package roller	40	oscillating equipment
69	scray	37	smoothing blade
67	cutter	34	epoxy resin drum
62	carpet guiding unit	33	heating
60	scray	29	heated drum housing
58	sewing unit	25	high-level tank for M.D.I.
57	carpet rolling-up unit	24	high-level tank for polyol
55	ventilator	23	high-level tank for polyol-chalk mixture
54	fixed infra-red field	20, 21, 22	pumps
53	movable infra-red field	18	M.D.I. storage tank
51	driving rolls for carpet	16	tank for polyol-chalk mixture (constantly stirred)
50	friction drive for drum	14	mill to reduce particle size
49	application of release agent	12	high power mixer
48	colour-paste metering (not shown)	11	scales
47	zeolite metering (not shown)	8	pump
46	high-level tanks for various additives	7	storage tank for polyol system
44	air metering (not shown)	6	screw for transporting chalk
42	metering unit	1	storage silo for chalk
41	mixers		

polyether or the filler. A high density foam can be reliably produced by the introduction of a controlled amount of air into the mixing head. A low density foam is produced by the addition of the correct amount of water, naturally without zeolite. In this formulation extra isocyanate is necessary to react with this added water.

The usual filler is natural chalk with a fixed particle size having a maximum water content between 0.05 to 0.15%. Barium sulphate and hydrated alumina are sometimes used to achieve special effects. The amount of chalk influences principally the viscosity of the reacting mixture, the density of a solid urethane coating and the price per unit weight of finished material. This price per unit weight is particularly important for loose-laid tiles because they need about 2½ kg of coating per m^2. The coating of a wall to wall carpeting is judged in a different way. A practically unfilled low density foam (density approximately 0.3) which is relatively expensive per unit weight can be preferable to a cheaper, more highly filled, high density foam (density approximately 0.5) because the weight of the coating per unit area can be considerably lower with the low density foam. This naturally assumes that the low density foam is just as or more suitable for the required purpose as the high density foam. Practically, other factors must be considered when compounding the coating, e.g., the influence of the viscosity (this depends on the amount and type of filler) on the penetration of the coating into the tufts, the filament and into the tufting substrate. A typical formulation for a solid coating with 62% filler is as follows:

100 p.b.w.	polyether
4 p.b.w.	zeolite paste (50%)
220 p.b.w.	chalk
31 p.b.w.	MDI

In this case the stoichiometric ratio NCO/OH is 1.15. The unfoamed finished coating has a density of 1.7 Mg/m^3. The density is reduced to 1 to 1.2 by the introduction of 50 to 80 cm^3 of air per kilogram of mixture into the mixing head.

A low density foam with 45% filler content is compounded as follows:

100 p.b.w.	polyether
0.3 p.b.w.	water
112 p.b.w.	chalk
38 p.b.w.	MDI

The density is 0.35.

Using 3 p.b.w. of silicone surfactant instead of water a mechanical foam (foamed with trapped air) can be produced by means of the well known latex foam type mixing plant. By varying the amount of air in the foam densities between 0.25 and 0.7 can be produced.

Properties of the Urethane Coating

We have had considerable experience over the years with urethane casting systems for other applications and this has helped us in working out the formulations, in recommending the most suitable processing conditions and in optimising the properties of the finished carpet backing. The well known elastic covering for running tracks and sports grounds in many stadiums of international reknown is based on a Bayer polyether urethane system. The urethane carpet backing has the same good resistance to weathering, against humidity and bacteriological attack. This can be an important factor in bathrooms, in cellars and when under-floor heating is installed. It is even possible to clean tiles with such a urethane coating in a washing machine. These urethanes do not contain any component which can be extracted by, for example, spilled perfume or stain remover, and the pile does not become stained in this way.

If it is wished to improve certain properties for special applications, it may be necessary to compromise and reduce the level of certain other properties. If, for example, a foam quality with good sound absorbing characteristics is produced, this can hardly be regarded as the best heat conducting type for under-floor heating. The so-called contract areas, for example offices and schools, require a particularly strong tuft anchoring. This can only be fulfilled with urethane if the foam backing is applied on to a urethane anchor coat. This pre-treatment can be accomplished on the same machine as one operation before the application of the foam backing. Several properties can be favourably modified by means of special additives, such as hydrated alumina. Modifications of this sort will play an increasingly more important role to help improve antistatic and flammability behaviour.

John Bunten

SATRA House
Rockingham Rd.
Kettering, Northants, England

PERFORMANCE REQUIREMENTS OF URETHANE-COATED FABRICS

(Received April 19, 1975)

INTRODUCTION

Within the last few years there has been a strong trend towards increasing use of materials other than leather for shoe uppers. In the long term this is an inevitable tendency, since the growth in the cattle population of the world is not as fast as the increase in world shoe production. Thus periodically leather is in short supply and therefore expensive. This encourages suppliers (of competing materials) to devise new materials, to increase their output, and to step up their sales effort.

As with any new material, faults and deficiencies become apparent in the early stages of use, and it is the job of the research institutes to lay down standards of performance to minimize the risk of unsatisfactory materials being used. This entails a considerable amount of work in the laboratory, in field trials, and in observation and monitoring of failures in shoes in actual full-scale production.

SETTING STANDARDS OF PERFORMANCE

The first step in setting a standard is to predict, or observe from actual practice, what types of failure can occur. Then a test or series of tests must be devised to reproduce the conditions causing the failure.

The next step is the most difficult and costly in time and effort – it is to validate the test, ensuring that there is a correlation between test results and incidence of failure. If no such correlation exists – and this situation is not uncommon – then we must go "back to the drawing-board", and try again. It is this step which has been omitted from a very large number of tests, often widely used, and the author does not know of any laboratory, including his own, which can honestly claim that all their test methods have been fully validated.

Having achieved a test method which proves to correspond to wear, then it is necessary to lay down levels of performance as a guide to show manufacturers and material suppliers alike. It is important to realize that only guide-lines are possible, not absolute pass-fail standards, although these may be used for convenience in certain cases. It is practically never possible to state categorically that a material is suitable for all classes of footwear, or for that matter that is completely unsuitable for all. The finest calf leather would fail quickly if used for heavy industrial boots. On the other hand, even paper can be used for some types of upper, and has, in fact, been made into "throw-away" shoes meant to be worn once only. Moreover, material failure in shoe manufacture can be the result of limitations in its properties, faulty manufacturing procedures (e.g. incorrect patterns, bad closing, too heavy lasting, excessive roughing, etc.) or combinations of these two aspects. Thus a report on a material is more likely to say "care is necessary in lasting", for instance, than "this material is unsuitable".

Similarly, wear failures may result in part from particular conditions of wear, so that the report will say "should not be used for footwear liable to be worn in cold climates", or "rather liable to scuffing damage and should not be used in children's shoes", rather than outright condemnation, as a rule.

URETHANE-COATED FABRICS

These are materials which consist of a thin urethane surface film laminated by means of an adhesive tie-coat to a woven, knitted or non-woven base fabric. Commonly the base fabric is raised, to give the required "handle" to the material, and this is the source of some of the problems. Unless the raised fibres form an even, dense pile, some of them will penetrate the top coat, affecting its appearance and durability. Also the weakest layer in the attachment of urethane to fabric is likely to be the strength of the pile.

Other variants of PUCF's include, as mentioned, alternative base fabrics such as non-woven materials or knitted cotton or nylon. The knitted materials are particularly suitable where high strength is important, such as in the legs of tight-fitting boots. Also foam PVC or PU has been used as an interlayer to give the required handle in place of the pile fabric. Not all of these have been successful, and by and large the same test criteria are applicable.

GUIDE-LINES FOR PROPERTIES REQUIRED OF PUCF'S FOR WOMEN'S SHOES

The following table gives representative figures for some of the important properties which can be used for guidance in the case of urethane-coated woven fabrics intended for use in women's shoes. As pointed out, it is often necessary to apply informed experience in interpreting the results of tests such as these. On the other hand, if a research institute is faced with the direct question – "Should I buy this material for shoes of this type?" – we must obviously be prepared to give a straight "Yes" or "No", based on our best judgement.

Some of the evidence on which the criteria given in the table are based is illustrated in the following photographs of failures encountered in the course of SATRA's laboratory testing work.

Urethane-Coated Fabrics for Women's Shoes – Guide Lines for Some Important Properties

Property	Requirements
1. Extension-at-break (tighter direction)	7%
2. Tear strength (weaker direction) initial maximum	3.0 kgf
3. Strength of Needle Perforations (for "seam" 3 mm from edge of test-piece) Weaker direction	1.6 kgf/cm
4. Lastometer Distension	6.0 mm
5. Peeling Load of Coating. Weaker direction	
Dry	1.2 kgf/cm
Wet	1.0 kgf/cm
6. Resistance to Flex Cracking	No cracks right through coating after:
At 20°C	500,000 flex
At –5°C	50,000 flex

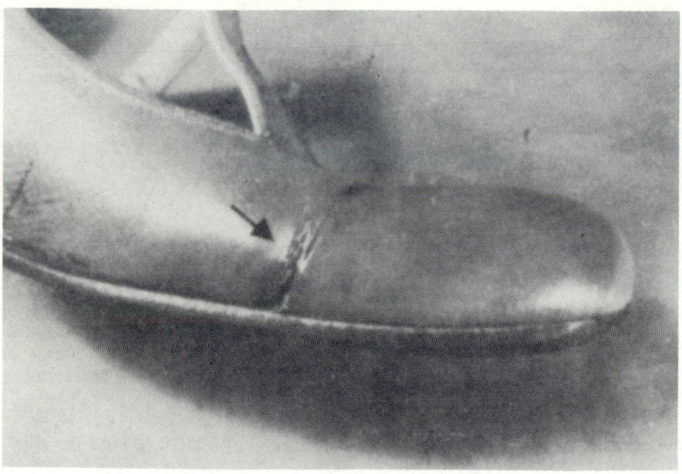

Figure 1. The low extension-at-break of this material (5%) has contributed substantially to the breakdown in wear.

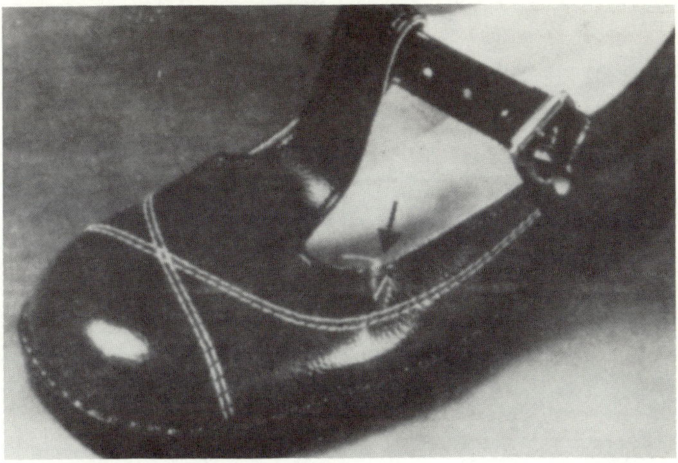

Figure 2. This is actually a PVC-coated fabric, but it illustrates clearly a torn top-line which could well occur equally with a PUCF. The initial tear strength of this material was 2 kgf.

Figure 3. A non-woven polyurethane-coated material which has burst around the feather. The tear strength was 3.0 kgf. which we would regard as borderline. Over-roughing, poor fit, or arduous conditions of wear may have played a part in the failure.

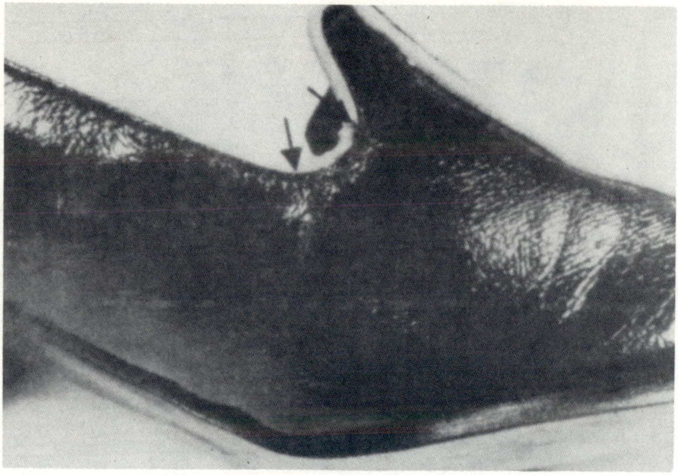

Figure 4. The double failure here is due to low initial tear strength (2.5 kgf) and low strength at needle perforations (1 kgf/cm).

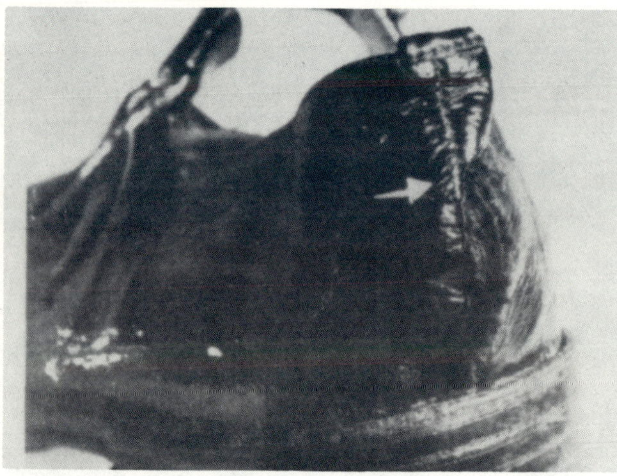

Figure 5. Again, the failure of the back seam is to be expected with 1 kgf/cm strength at needle perforations.

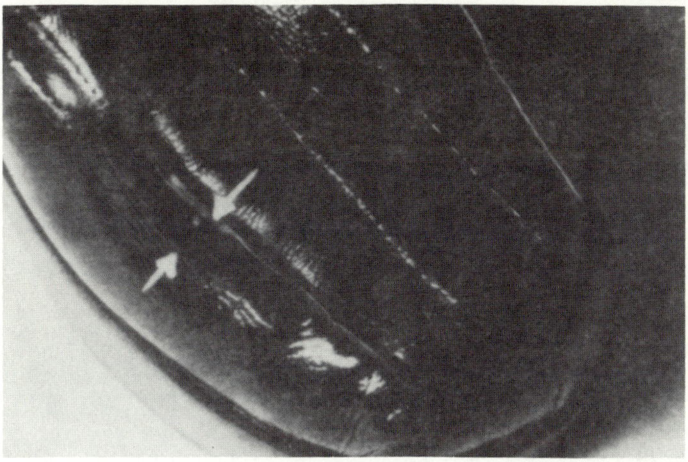

Figure 6. Complete failure of the material at the seam along the shoe is accounted for by a strength at needle perforations of only 0.5 kgf/cm.

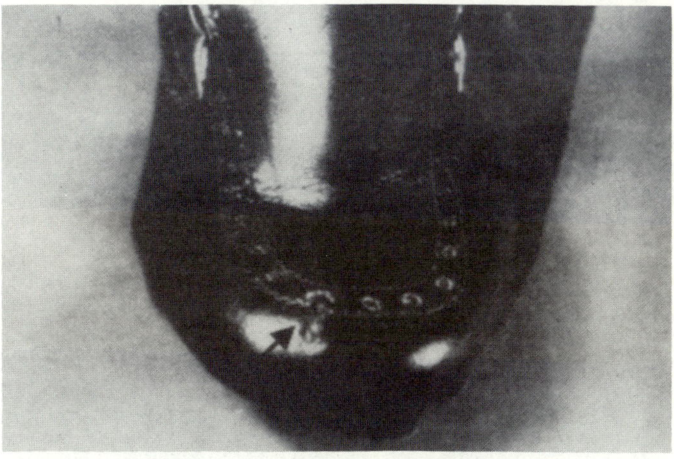

Figure 7. The low lastometer distension (5.5 mm) accounts for these breaks which occurred during lasting.

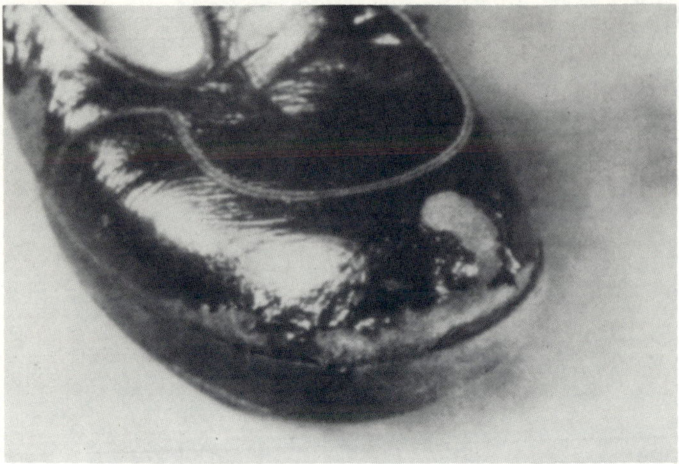

Figure 8. Low peel strength showing inadequate adhesion of the top film is a common fault of PUCF's. This one gave 0.6 kgf/cm dry.

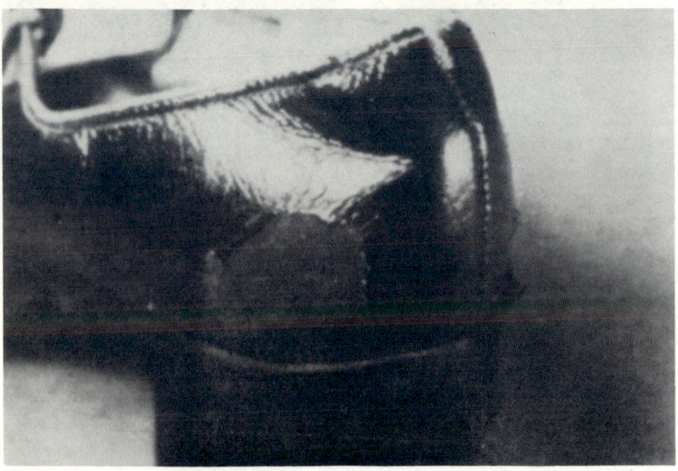

Figure 9. And this 0.4 kgf/cm.

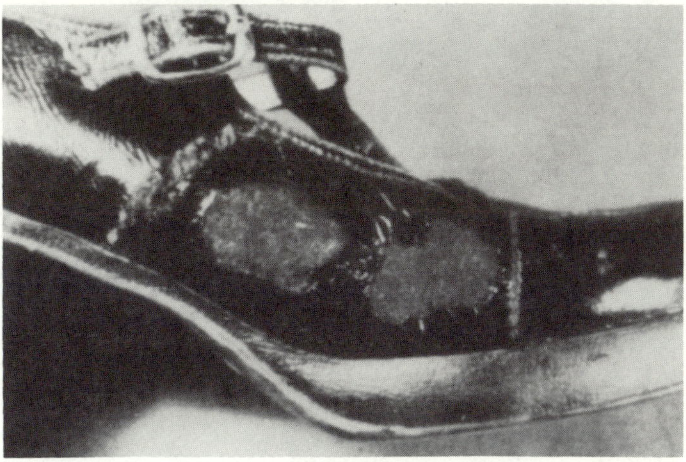

Figure 10. The double failure here is due to low peel strength (0.6 kgf/cm) and low strength at needle perforations (1 kgf/cm).

Figure 11. Low wet peel strength is the cause of this failure 0.4 kgf/cm.

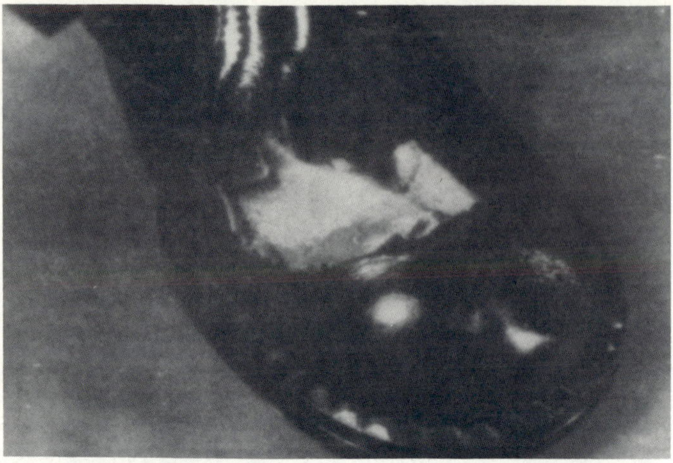

Figure 12. This shows a very brittle top-film on a PUCF, which gave only 100,000 flexes at room temperature in the flex cracking test.

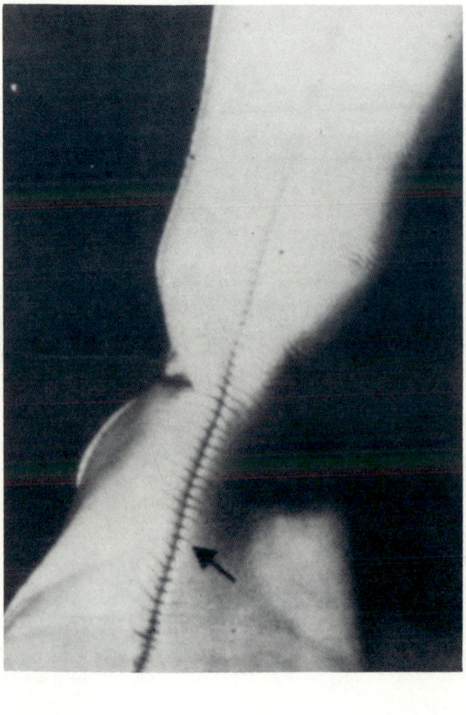

Figure 13. The failure along the seam of this polyurethane coated leather is due to a borderline flex cracking resistance – 500,000 flexes – together with a boot design which calls for better-than-usual flex resistance.

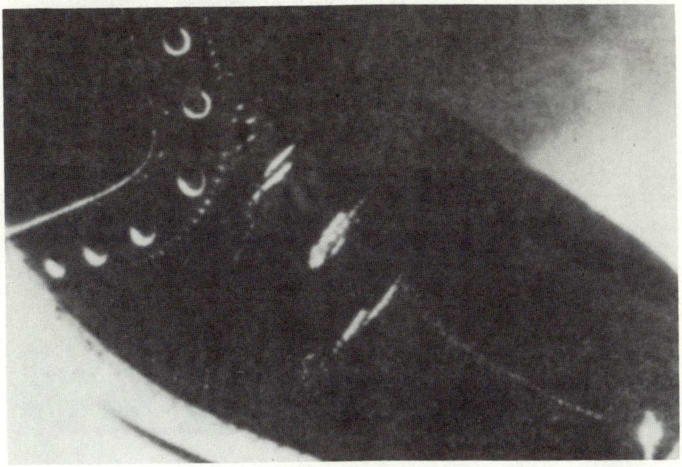

Figure 14. Again a PU coated leather whose weakness was in wet flexing (10,000 flexes).

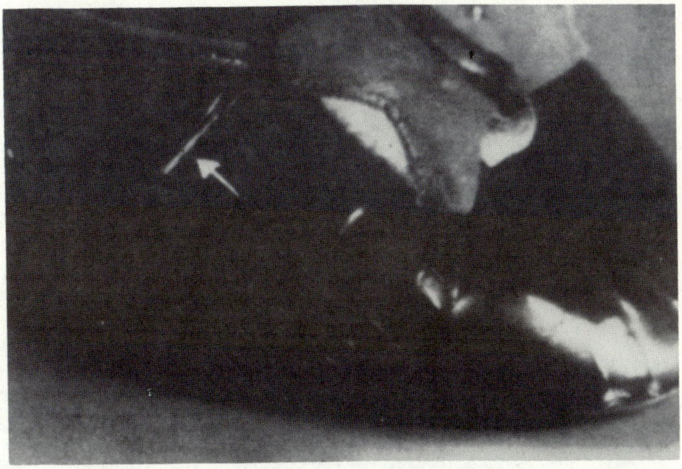

Figure 15. This PU-coated leather gave only 100,000 flexes in the flex cracking test.

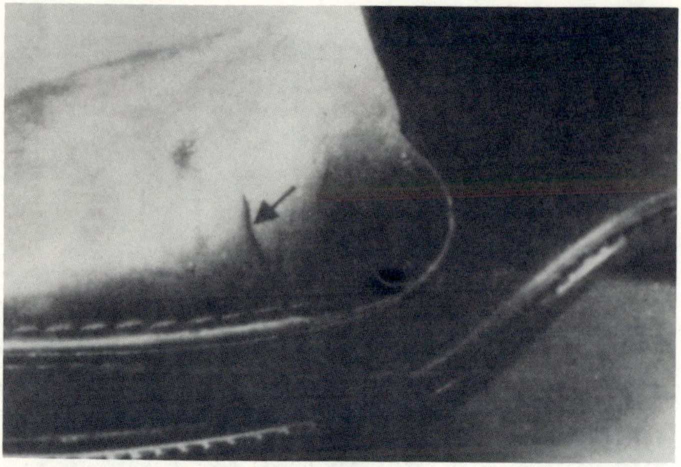

Figure 16. This flow moulded PVC-coated fabric gave 500,000 flexes in the flex-cracking test. Footwear of the type these materials are commonly used for are expected to stand up to fairly hard wear, and our standard is 1 million flexes.

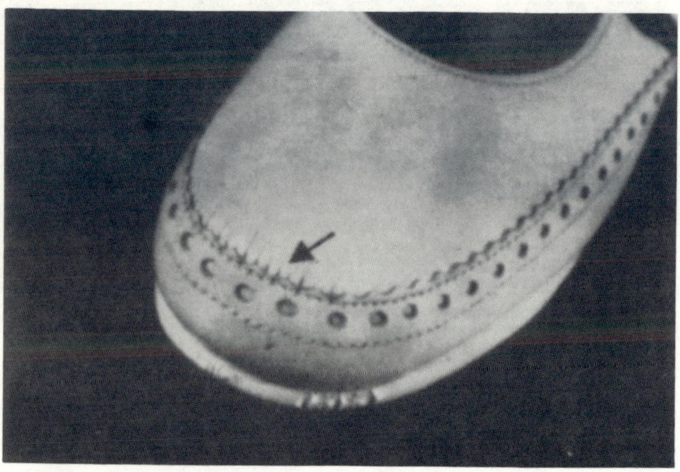

Figure 17. Some materials are found to fail during heat setting. The test we use is to stretch the material on a Zwick tester (with and without a row of needle holes) and then to apply a heat setting treatment. This material is a PU-coated leather.

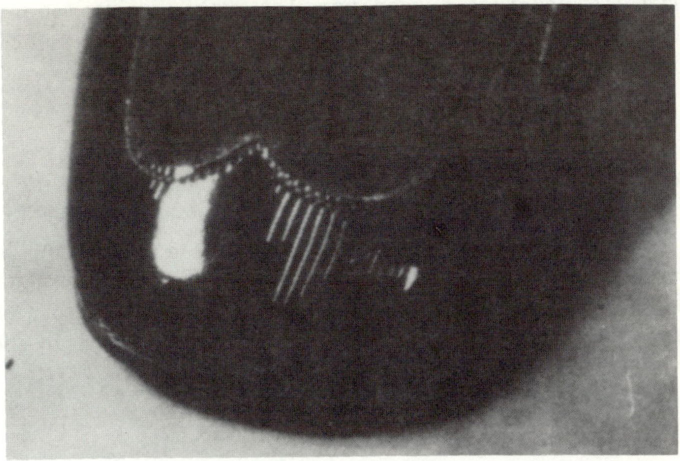

Figure 18. Another example of cracking caused by heat setting which can be predicted by the same procedure using the Zwick tester.

Figure 19. Another rather spectacular failure of the same type.

We have not succeeded so far in devising a form of snag/scuff test which reflects the type of failure most commonly met in PUCF's and which is also suitable for publishing as a standard recommendation. The method we use can usually predict the tendency to snag, but is not easy to define in such a way that other laboratories could reproduce the same results.

CONCLUSION

This note describes the procedures which have to be adopted in the evaluation of one type of upper material.

When a new type of material is being developed by a supplier, intelligent guesses have to be made as to likely modes of failure, in order to help the supplier with the development. These are supplemented progressively by laboratory tests, making trials in the shoe factory, (small scale) wear trials and bulk trials by shoe manufacturers.

ACKNOWLEDGMENTS

To Richard Turner, at SATRA, who is responsible for routine testing of upper materials and to Geoffrey Hole, at SATRA, Head of Materials Research Department.

L. G. Hole and P. D. A. Lacy

Shoe and Allied Trades Research Association
SATRA House
Rockingham Rd.
Kettering, Northants NN16 9JH, England

TRANSFER PRINTING ON URETHANE COATED FABRICS

(Received September 16, 1975)

HISTORICAL BACKGROUND

Transfer printed materials have been available for at least a century. The earliest embroidery transfers were produced from pigmented shellac inks, applied by hot iron to linen or cotton. More recently, pigmented lacquer prints have been produced on release paper for transfer under heat and pressure to plain materials, but this type of decoration is usually confined to motifs because the transferred film affects the handle of the substrate.

The advance in the production of synthetic polymer textiles that has occurred in the last 20 years has had a marked effect on the dyeing methods used. The traditional copper-roller has been superseded by flat-bed and rotary screen machines but the commercial introduction of a vapour phase transfer method, by the French firm Filatures, Prouvest, Masurel La Lainiere de Roubaix in the late 1960's can be said to have had a dramatic effect on synthetic textile printing. Estimates of the use of the technique indicate a growth to 100 million square yards in 1972 [1].

DESCRIPTION OF VAPOUR-PHASE PROCESS

The transfer method, known as vapour-phase, dry-heat, sublimation or "Sublistatic" is one whereby a pre-printed design using specially formulated inks containing sublimable, disperse dyestuffs are produced on a paper support. The dye image is then transferred via its vapour phase to a dye receptive, synthetic fabric under controlled conditions of time, temperature and pressure.

Printing of the transfer paper can be carried out by conventional methods used in the graphics' industry. At present gravure, lithographic, flexographic, letterpress and screen techniques are used. Each printing process offers a particular blend of cost, print quality and suitability for a particular market. The largest proportion of papers available at present are printed on the roll by gravure. This gives, the best reproduction with virtually no design limitations and print consistency over very long runs. In comparison, screen printed papers are produced only in sheet form and have some design limitations but have an obvious advantage for short runs (of exclusive designs) because of the relatively low preparation costs.

An adequate range of disperse dyes that sublime in the region of 180°C are available, which are readily absorbed in the vapour phase by most synthetic polymer fibres. Azo and Anthraquinone dyes are amongst those in use, selected for their vaporisation rates and substantivity for a particular fibre. The molecular weight range of the majority of disperse dyes used is 240–340. An intensive study of the likely classes of dyes has been carried out by dyestuff suppliers and transfer paper manufacturers. Burtonshaw [2] describes the effect of molecular substilation on sublimation rates of disperse Azo and anthraquinone dyestuffs (Figure 1).

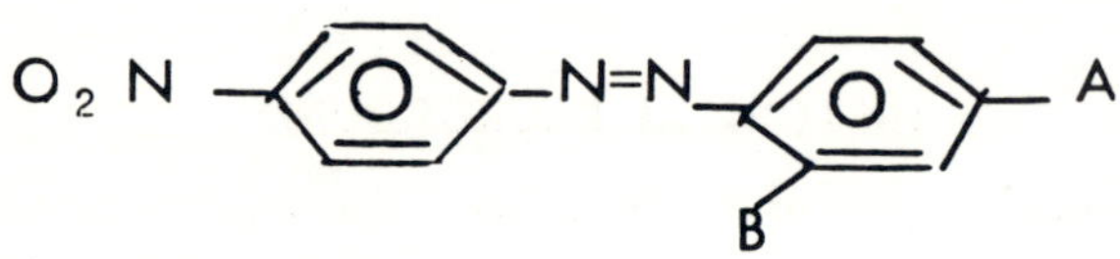

Azo dye substituents :

A = OH, NH_2

B = OH, NH_2

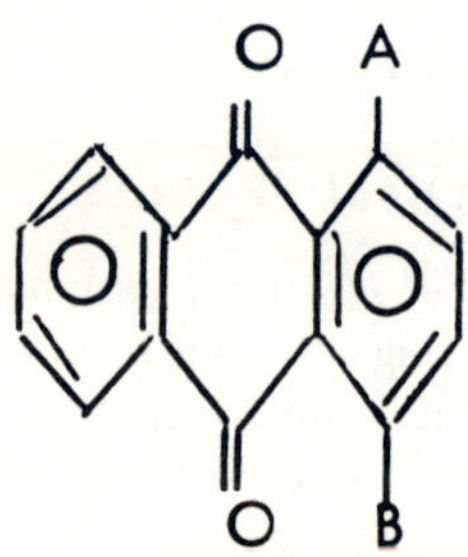

Anthraquinone dye substituents :

A = NH_2, $NHCH_3$, NHC_2H_4OH

B = NH_2, $NHCH_3$, NHC_2H_4OH

Figure 1. Examples of Azo and anthraquinone disperse dyestuffs.

In both dyes an increase in molecular size caused by substitution, and the presence of a terminal hydroxyl group adversely affected the dye transferrate by sublimation.

In a similar review, Holland and Netherland [3] detail the effect of molecular size and structure on dye fastness properties when transferred onto a range of synthetic fibres and also recommend transfer conditions for each fibre. The performance of this class of dyestuffs on fabrics is therefore well documented, both with respect to transfer rates and subsequent print performance. The printing process is a simple dry heat transfer which takes place when the transfer paper and fabric are held face to face in flat-bed or continuous, heated roller press. The dyes vaporise, are deposited on the outside of the fibres and then migrate into them. The printed fabric requires no finishing treatment and, provided the dyes are correctly selected, high fastness properties are achieved.

The following are among the advantages claimed for transfer printing of fabrics in comparison to conventional methods:

Simplicity:	a clean, dry heat process
Design:	virtually any intricate multicolour design can be reproduced with outstanding definition and no registration problems. A design printed onto different colour backgrounds gives a range of effects.
Flexibility:	plain fabric is converted only after sales trends have been identified.
Low stockholding cost:	paper replaces printed fabric as the stock material and hence reduces stock value.
Low spoilage:	print faults are overcome at the paper production stage

In comparison there are a few disadvantages, which include:

Colour range:	a more limited colour range can be achieved than by conventional methods, as there are some shades difficult to match using sublimable dyes.
Fastness:	the dyes have limited fastness properties with some fibres, e.g. nylon, but difficulties can be minimised with careful selection.
Urgent colour changes:	no last minute change of colour is possible, once the papers are printed.

From the rapid growth of the heat transfer process, it would appear that for the textile market the advantages claimed have been found realistic.

TRANSFER PRINTING OF POLYMER FILMS

A number of the advantages listed for heat transfer printing of textiles would be valid for the decoration of polymer films, for example PVC and polyurethane sheet or coated fabric.

At present these materials can be printed by conventional methods and special spraying effects can also be used. Gillibrand [4] has described the aesthetics of coated fabrics and the conventional methods that are available for decorating them.

In recent years attempts have been made to use the heat transfer process to print polymer films using conventional sublimable dyes selected for textiles, but these have generally been unsuccessful for well defined reasons:

i. The dyes are soluble in the low molecular weight phthalate plasticisers used in flexible PVC coatings and are not completely fast in most polyurethane elastomers (and many other common coating polymers). Prints on these materials therefore, show a pronounced loss of definition in a few days or weeks, depending on the ambient temperature.

ii. The dyes migrate into other polymer films when in contact with them. Again, heat accelerates this process.

iii. The light fastness of dyes is generally lower in PVC and polyurethane films than in synthetic textiles and performance is unpredictable. Hence a typical disperse blue dye will (eventually) fade in sunlight when printed on polyester fabric but will change shade (to red) when used in a commercial thermoplastic urethane coating.

In 1973 SATRA investigated the use of transfer printing shoe upper materials. The main objectives were to overcome the known shortcomings and to establish optimum transfer conditions for printing thin thermoplastic polymer films without damage to the polymer or (in the case of coated fabrics) the material construction. A joint development programme was initiated with Alf Cooke Limited, Crown Point, Hunslet Rd., Leeds, a member of the Bemrose Corporation, a major supplier of heat transfer papers to the textile industry.

Transfer Conditions for Polyurethane Coated Fabrics

Textile transfer printing depends on the ability of the selected disperse dyes to sublime, generally in the temperature range 180–210°C. When these same dyes were used to print polyurethane coated fabrics high colour transfer occurred at 180°C but the polymer film was irreversibly damaged. Transfer conditions in the range 110–180°C, with dwell times of 30–180 secs. were examined [5] to determine the degree of damage to the polyurethane top film, the tie-coat polymer and hence the physical properties of the material. Tensile breaking properties and film to fabric laminar strength were monitored, whilst scanning electron micrographs of coated fabric cross-sections provided additional information.

The following results were obtained with a white commercial polyurethane coated fabric (Figure 2). The effect of print time and temperature on the breaking elongation are shown in Figure 3 whilst the electron micrographs of Figure 4 clearly show the corresponding polymer film deterioration as the print temperature is

Dwell temperature (°C)	FILM PROPERTIES		Laminar strength (Peel) (Kgf/cm) Transfer time = 60 seconds
	Tensile breaking load (MN/M^2) Transfer time = 30 seconds	Breaking elongation (%) Transfer time = 30 seconds	
Original material	9.90	351	1.00
110	10.72	359	1.00
130	10.40	345	0.94
150	9.47	335	0.91
170	9.04	295	0.86
180	8.90	290	0.82

Figure 2. Effect of transfer temperature on tensile and laminar strength.

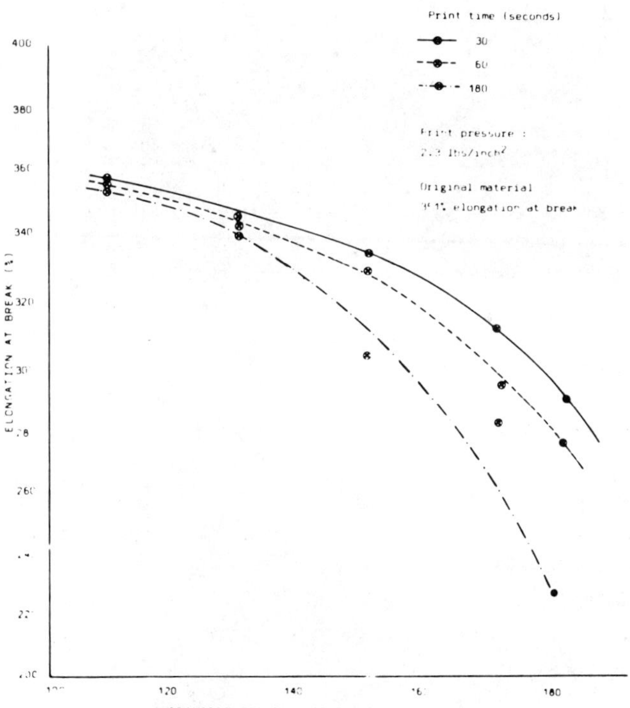

Figure 3. Effect of print time and temperature on tensile-breaking elongation.

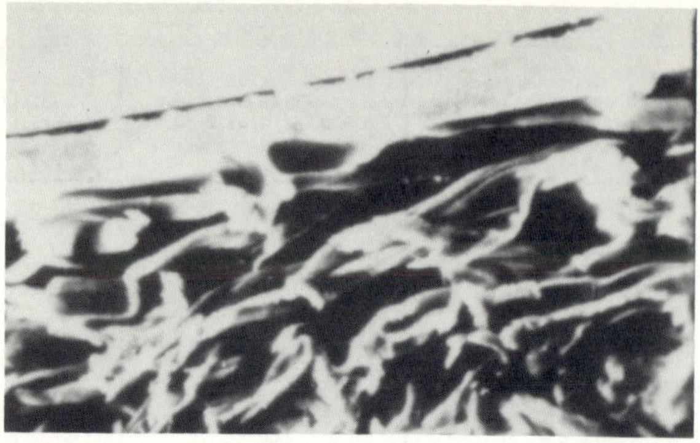

(A) 220×

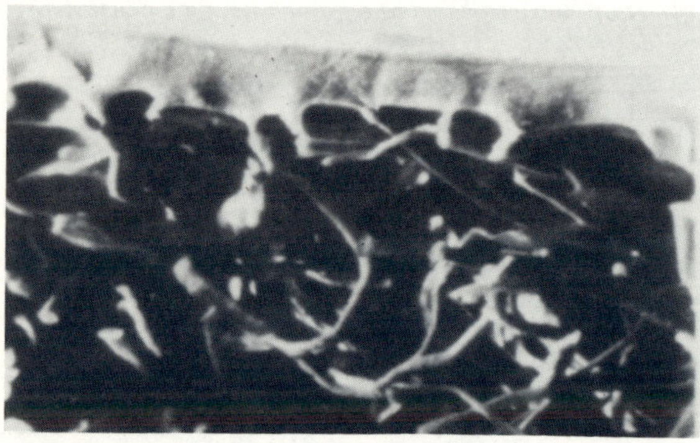

(B) 220×

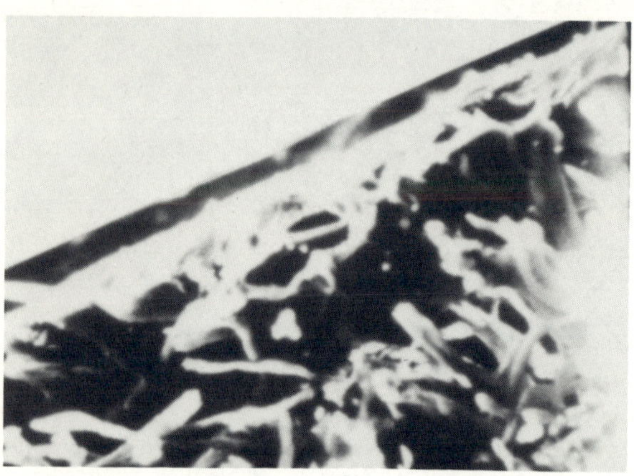

(C) 220×

Figure 4. Scanning electron migrographs of polyurethane coated fabric transfer printed for 60 seconds at (a) 130° C, (b) 150° C and (c) 180° C.

increased. Clearly, the transfer temperature must not exceed 150°C, with a dwell time of 30 seconds.

Efficiency of Dye Transfer to Polyurethane Films

Having established the effect of transfer temperature and dwell time on the physical properties of polyurethane coated fabric the efficiency of a blue anthraquinone disperse dye transfer was investigated under the same conditions. The dye was soxhlet extracted from the transfer paper, using methyl ethyl ketone solvent, before and after printing. Dye concentrations were estimated from their photometric absorbance at 634mm in the visible regions and are given in Figure 5, shown graphically in Figure 6.

Transfer temperature (°C)	Transfer time (secs)	Efficiency of dye transfer (%)
110	30	40
	60	48
	120	63
	180	65
130	30	65
	60	78
	120	83
	180	89
150	30	83
	60	88
	120	94
	180	94
170	30	93
	60	97
	120	99
	180	99

Figure 5. Contact time and temperature effects on percent dye transfer.

Visual examination of transfer printed patterns indicated that the colour transfer to polyurethane film should be 70–80%, depending on the actual dyestuffs selected, and the limit conditions imposed about (150°C, 30 seconds dwell) were therefore shown to be acceptable.

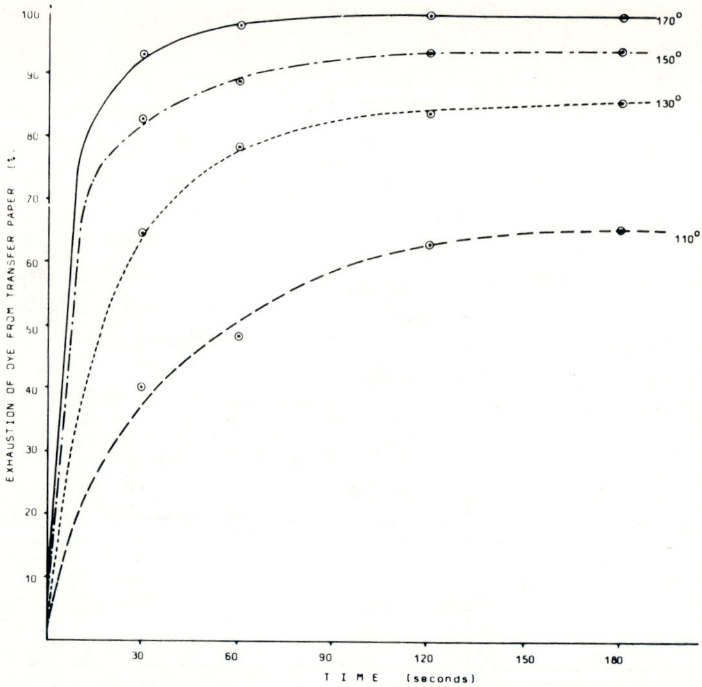

Figure 6. Effect of contact time and temperature on blue anthraquinone dye transfer onto polyurethane.

It was interesting to note that the disperse dyes did not transfer to polyurethane by sublimation alone as previous work [2], on textiles, had shown that little sublimation occurred below 150°C, with the majority of dyes in use (Figure 7). Further work at SATRA established that solid phase diffusion was the dye transfer mechanism for polymer films of this type, and therefore required good contact of polymer and transfer paper. This was dramatically illustrated when a transfer press with platen surfaces out of alignment was used, with disastrous results (Figure 8). The arrows indicate poor colour transfer corresponding to a recessed bolt-head in the heated platen [1] and general platen misalignment [2].

Dye Migration in Polymer Films

Since the transfer process is a time and temperature dependent diffusion phenomenon it is not unexpected that at room temperature dye migration will aslo be observed but over longer time spans. We have shown that migration of dye from printed to unprinted polyurethane film is generally unacceptable when tested at 4 hours at 60°C (BS/3662/6 Colour Fastness to Marking-off). At the same temperature, migration within the polymer film is noticeable within 24 hours, and is

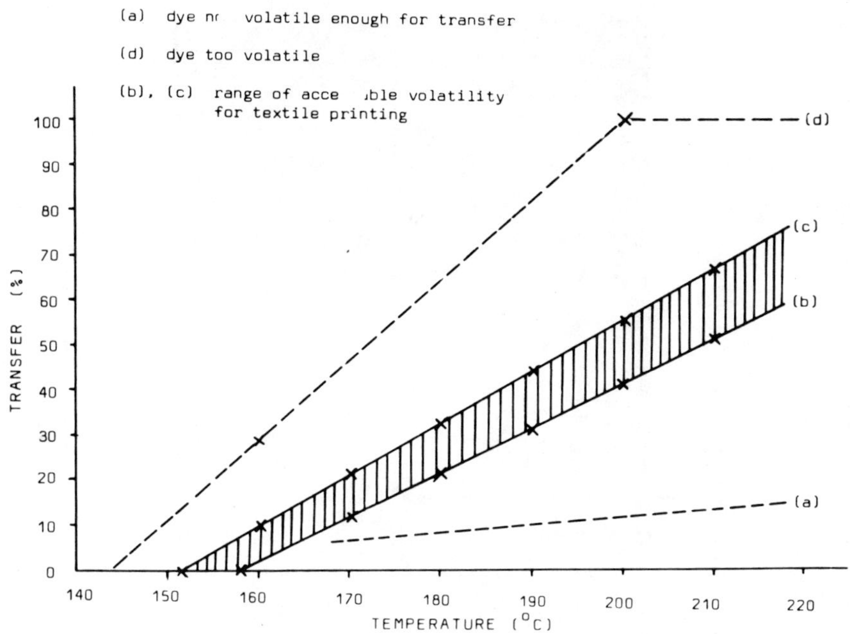

Figure 7. Effect of temperature on disperse dye transfer by sublimation to fabrics.

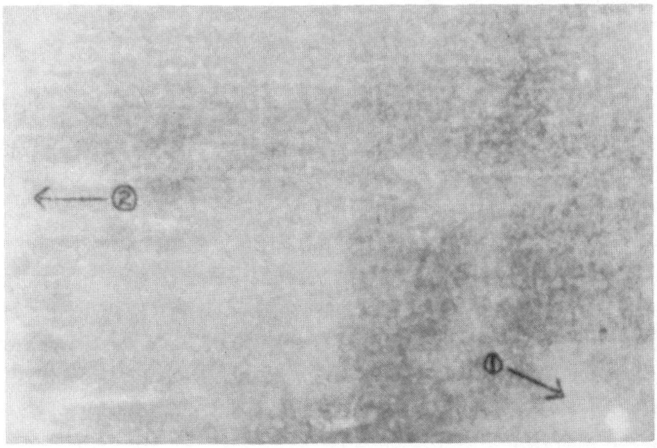

Figure 8. Illustration of solid phase dye transfer, with platen press plates out of alignment.

visible in a few days even at room temperature with most polyurethane coated fabrics. Migration is faster in flexible PVC films, where the plasticiser acts as a good

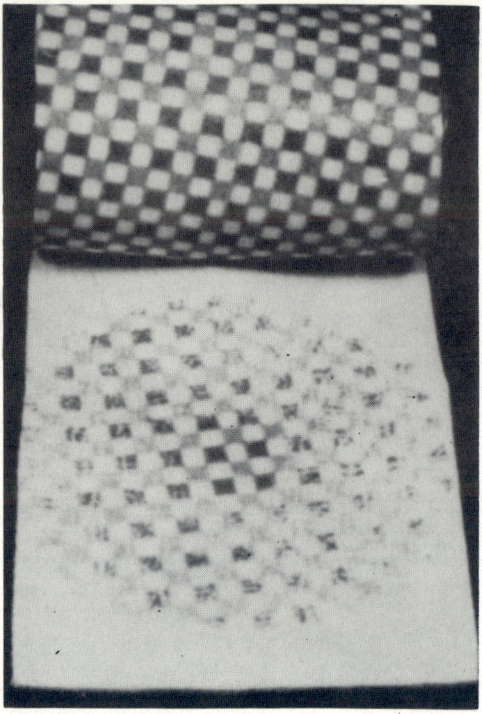

Figure 9. Dye "mark-off" from printed to unprinted PUCF after 4 hours at 60° C.

solvent for the dyes, and also occurs to some extent with all the other footwear materials (poromerics, leather, solings) that we have examined. Dye migration within polymer films can be estimated with a travelling microscope using the square print shown in Figures 9 and 10.

Migration is characterised by a loss in colour intensity of the print in addition to the blurring of edges. As it is a temperature dependent process great care is needed in the measurement of the colour fastness of transfer printed polymer films. If an artificial weathering cabinet is used the temperature of the sample will affect both the exposed and shielded sections of the material, leading to a false assessment if heat migration has occurred.

A secondary, but still important, result of the diffusion transfer mechanism is the difference in colour intensity found when the same transfer paper is used to print different materials, even of the same class. For example (Figure 11), most commercial, white PUCF's appear different when printed with the same transfer paper, because the dyes can penetrate further into the materials with thicker polyurethane coatings. The presence of a clear finish lacquer also affects the apparent depth of the printed pattern.

Figure 10. "Shop-window" heat-migration test (24 hours at 60°C) shows diffusion of printed pattern (bottom).

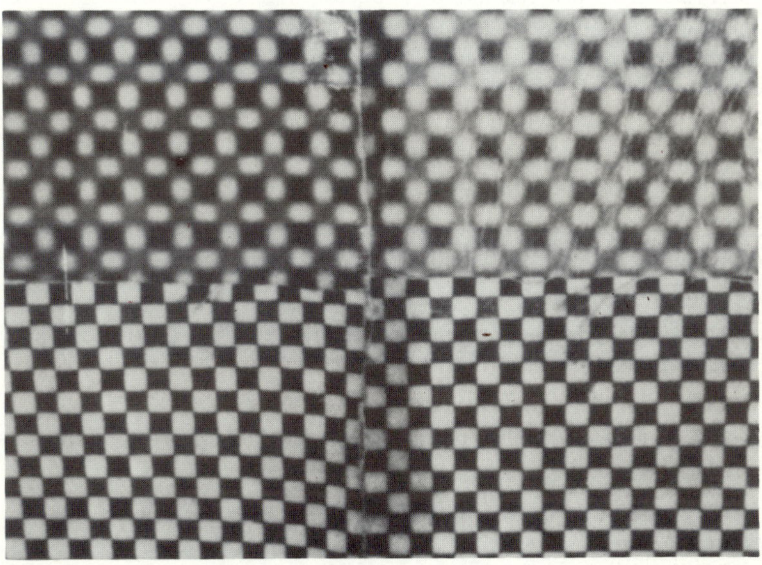

Figure 11. Difference in colour intensity of transfer print on two commercial white PU coated fabrics, top – unlacquered, bottom – lacquered.

Attempts to find dyes with suitable transfer properties which would not migrate when printed failed and it became clear that a means of stabilising the dyes was essential if the process was to become commercially viable. The method developed at SATRA provides a means of obtaining much improved colour intensity, eliminating the differenes between base materials. The SATRAPRINT[1] process eliminates dye diffusion and mark-off problems and marginally improves the light fastness of range of disperse dyes specially selected for their substantivity for the polymer.

The material to be used is prepared by coating it with a special lacquer having high affinity for the dye. If a coated fabric is to be printed, the special coating can either replace the existing finish or can be applied as a top lacquer. Dye migration properties are dependent on this lacquer coating thickness but at the optimum film thickness dye diffusion appears to be completely eliminated when used with almost all polyurethane coated fabrics and other urethane based upper materials. Test samples have been aged 6 months at 60°C without a noticeable change in the printed pattern.

Evaluation of printed polyurethane coated fabric upper materials has shown that film peel strength and flex resistance are unimpaired and that the abrasion resistance of the print is excellent due to its penetration into the coating. Hydrolysis resistance of printed polyurethane films does not appear to be affected by the presence of the disperse dyes although the special coating reduces water vapour permeability and the gloss level of the polymer film is slightly reduced by the printing process.

SCOPE FOR NEW FASHIONS IN SHOES AND ASSOCIATED FASHION WEAR

The SATRAPRINT process may be applied to rolls of PU coated fabrics and for such applications as upholstery, garments and other large quantity applications, this might be the preferred way. Transfer presses are readily available of suitable size, either in platen or roller form. The latter has obvious advantages if an overall pattern is to be printed on to material rolls by the manufacturer or on a commission basis. For the shoe manufacturer, however, there is much in favour of a process where individual upper components are printed in his own factory; this has at least three advantages:

i. Assist with regard to short production runs with a particular transfer paper pattern and colourway.

ii. The freedom given to the imaginative designer, who can create a new exclusive appearance for an upper without the restriction of using the currently available colour ranges supplied by material manufacturers.

iii. White PU coated fabric only would be stocked - all colour-pattern effects would be obtained by printing.

[1] SATRAPRINT Trade-Mark and Patents applied for

iv. Minimise paper costs.

Obviously the printing process will add to shoe upper production costs, but economies are possible as follows:

i. A number of colourways would be available with a particular design. Once the plate costs had been met the expense of a different colourway would be minimal. For example, a colour range of polka-dot design would be obtained from the one plate (it is assumed that a lithographic or similar batch paper printing process would be essential because of the short run requirement, especially for shoes).

ii. One plate can be made to carry several individual designs provided that the colours are compatible. Sheets printed in this way may then be stacked and cut so as to produce sets of small sheets of one design and of a size to print one upper. Thus one printing plate could produce six or eight different exclusive patterns.

iii. A suitable mono-colour pattern could be printed onto a range of PU coated fabrics of different pale colours. The base material colour would form the background for example in the unprinted areas of gingham pattern. Thus three mono-colourways on the transfer paper and three PU coated fabrics - one white and two pale colours - would provide nine different colourways on the printed materials.

One other development of the SATRAPRINT process is its co-ordination with the HF flow-mould technique. It is relatively easy to incorporate printed areas on a flow moulded upper and this is exemplified in Figure 12. This figure is an HF emobssed PU coated fabric, not a flow-moulded PVC, and the print decoration coupled with the HF embossed grain has created an interesting effect. Both the transfer printing and the embossing takes place simultaneously in the HF press.

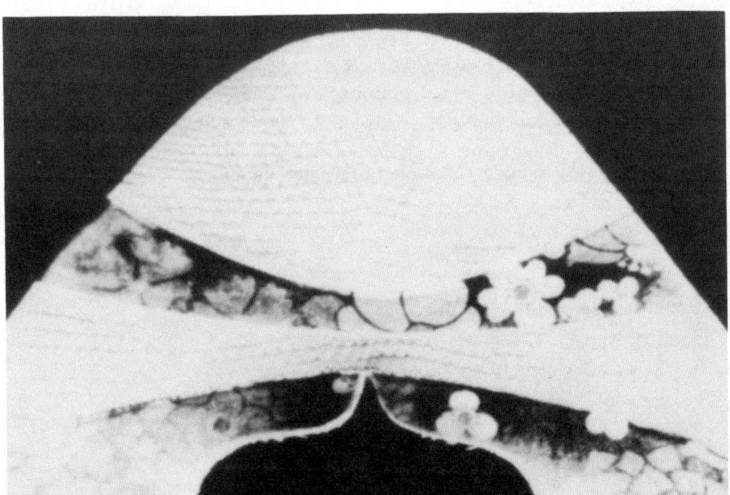

Figure 12. Ladies PU coated fabric upper printed and flow-moulded using HF.

Figure 13. Examples of PU coated fabric shoes decorated using the SATRAPRINT process.

The scope for fashion and co-ordinate shoes - garments, accessories, would seem to be simply awaiting the creative powers of designers and the fashion houses. Some typical shoes decorated by the SATRAPRINT process are shown in Figure 13, much is lost due to the reproduction being in black and white rather than in colour.

REFERENCES

1. C. E. Vellins, 6th Shirley International Seminar 1973.
2. D. W. H. Burtonshaw, Textile Journal/Australia, December 1972, 16–26.
3. G. Holland and A. Litherland, JSDC, December 1971, 488–495.
4. Gillibrand, 4th Congress, Scientific Society of the Leather, Shoe and Allied Industry, Budapest.
5. M. T. Kasapoglu, MSC Project 1973, Loughborough University.

JOHN R. DAMEWOOD

Polyurethane Specialties Division
Millmaster Onyx Corporation
Lyndhurst, N.J. 07071

URETHANE TRANSFER COATING*

(Received May 29, 1975)

INTRODUCTION

THE CASTING-ON-PAPER (better known today as *transfer coating*) process first appeared on the coating fabric scene about 1959 when expanded vinyls became commercially available. It was the middle to late sixties before urethanes (with their inherently more difficult coating problems) began to take their proper place in transfer coating. The last eight to ten years of urethane technology has produced remarkable achievements and depth of penetration in the coated fabrics industry. Exciting and rapid advances in the areas of polymer design, performance and aesthetics have led to the use of urethane coated fabrics in such demanding and various areas as men's and women's shoes, upholstery, wearing apparel and accessories such as handbags and billfolds. However, during this period, with the exception of advances in transfer paper technology, most of the techniques involved in urethane transfer coating have been independently developed by the numerous companies involved in transfer coating. This contribution presents some of the techniques, ideas, theories and postulations the author has been exposed to during his career in the coated fabrics industry.

Following are the critical areas involved in the production of a good quality transfer coated fabric. Each of these areas will be discussed in detail.

1. Formulation, blending and mixing
2. Machine and paper alignment
3. Skincoat coating station design
4. Skincoat drying oven
5. Adhesive coating station
6. Fabric Lamination
7. Adhesive drying oven
8. Separation of composite and release paper
9. Post finishing operation

In addition, a short discussion of the advantages and disadvantages of light stable, aromatic, thermoplastic and crosslinking urethanes will be included.

Choice of Polymers

In addition to learning the techniques of transfer coating, the coater must also

*Paper originally presented at Seminar on Urethane Coated Fabrics, May 29, 1975, sponsored by the Program Division, Technomic Publishing Co., Inc., Westport, CT. 06880

have a basic working knowledge of urethane chemistry. Questions such as the following must be answered before a formulation can be prepared. Should an aromatic or light stable polymer be used? Would a thermoplastic or crosslinked system be the better choice? What systems give increased solvent resistance? Some of these properties are shown in Table 1. The coater must now decide which system to use in his formulation.

Table 1. Properties of Urethanes

Aromatic urethanes

1. Lower cost
2. Solvent resistance
3. High tensile strength

Aliphatic urethanes

1. Light stability
2. Hydrolytic stability
3. Abrasion resistance

Crosslinked systems

1. Stain resistance
2. Possibly better adhesion to the fabric

Finished products, such as wearing apparel and shoes, must by necessity use light stable materials; however, an aromatic adhesive can be incorporated into the composite. Resistance by the coating to delaminate in the presence of solvents is increased when an aromatic urethane is used. With the exception of pastel colors, there appears to be no reason why the lower costing aromatics cannot be used for pocketbooks, billfolds and other similar accessory items.

Generally speaking, for most transfer coated products, light stable thermoplastic resins are used for skincoats, while light stable, aromatic and crosslinked systems are used for the adhesive coat. One factor that must not be overlooked in the choice of urethanes is the importance of the inter-relationship of each component (skincoat, adhesive coat, fabric) to the finished product. The flex life, abrasion resistance, etc. are all dependent upon the marriage of each component and not upon a single factor.

Formulation, Blending and Mixing

Most solution urethanes, whether thermoplastic, heat-activated or two-component crosslinking systems readily accept additives such as pigments, flatting agents or silicones. The primary criteria being the addition of only the minimum amount of additives necessary to achieve the desired end product aesthetics. The reasons for this are obvious; first, there is a cost factor involved, and second, the addition of excessive additives decreases the physical strength of the urethane. This is clearly illustrated in Figure 1. It is also advisable, although not necessary, to use pigments which have been milled in a compatible base resin. In the past and even today vinyl pigments have been used with urethane resins. Due to the possibility of plasticizer migration in the urethane (perhaps causing unsightly spots or splotches on the surface of the finished product, or isolated resin stiffness resulting in cracking) this procedure is not generally recommended. The best of all worlds is achieved when the pigment is milled with a low molecular weight polymer that contains a similar, if not the same, backbone structure contained in the urethane. This technique guarantees compatibility between the pigment and the urethane.

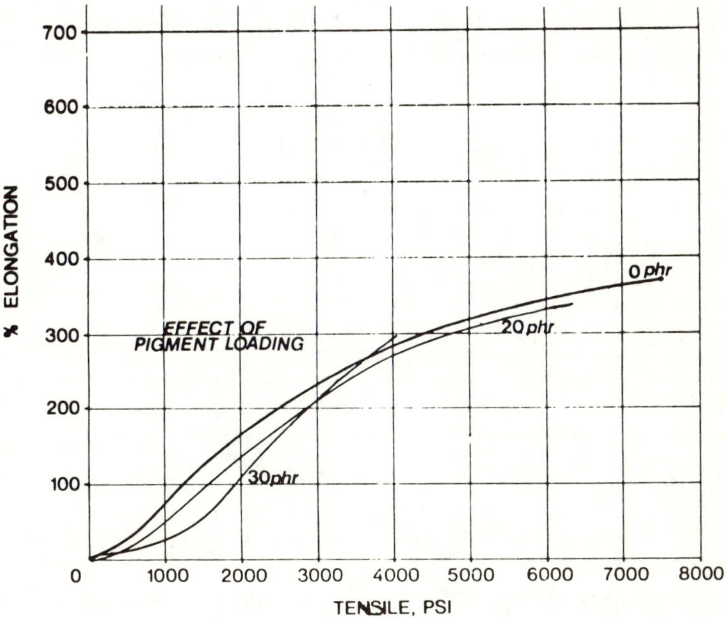

Figure 1. Effect of pigment loading on physical properties.

Normally an 8–12% pigment concentration (based on solids content) is sufficient for good coverage; however, in the case of pastels, particularly yellows, a higher pigment level is necessary, and even pigmentation of the adhesive coat is sometimes required for adequate hiding power. Generally, silicones are added for slip and leveling (even flow of the coating onto the release paper) in amounts varying from less than one percent up to a maximum of approximately five percent. These materials enhance the slip and aesthetics of the end product, in addition to aiding the release of the coating from the transfer paper.

Flatting agents such as silica, talc, etc. are usually used in the topcoat or printing operation to control the luster and degree of dryness in the final product; as much as fifteen to twenty percent has been used in some formulations, but such high levels are not normally recommended. The quantities used are obviously dependent upon the finished product requirements, i.e. shoe upper, wearing apparel, upholstery, etc. Each requires its own special formulations and there is even a wide range of variety within each group.

Prior to coating, all of the ingredients are combined and thoroughly mixed until a homogenous blend is obtained (usually 15–30 minutes). Mixing should be performed with a good Oaks, or similar mixing apparatus; however, the coater should be careful not to use equipment that is over or under engineered for the job. As shown in Figure 2, the polymer viscosity is dependent upon the mixing shear rate. Following the blending operation, and prior to coating the compound should be checked for color (compared to a standard) and the film properties.

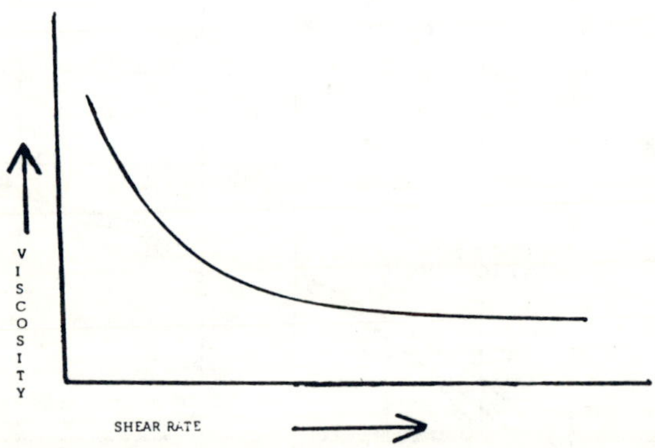

Figure 2. Effect of shear on viscosity.

Figure 3 illustrates a typical tensile modulus diagram of a urethane used as a skincoat. Materials in this general range exhibit good elongation, recovery and resistance to abrasion. In combination with the proper adhesive and topcoat, polymers in this group are excellent for shoes, upholstery and wearing apparel. With the exception of deaeration and filtration (which is discussed in the section on skincoat coating station design), the coater is now ready to move on to the operation of his equipment.

Machine and Paper Alignment

The importance of proper machine and paper alignment cannot be overemphasized. Without proper alignment, not only will the coater not be able to

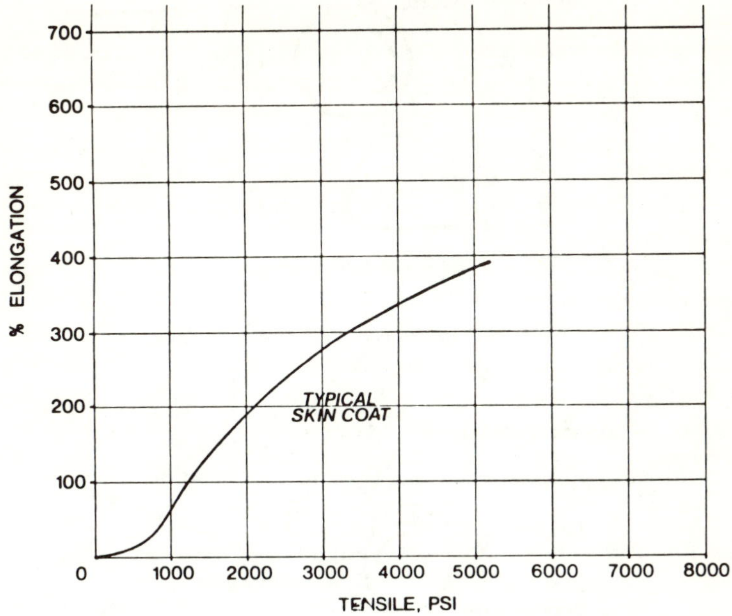

Figure 3. Typical tensile modulus diagram of a urethane used as a skincoat.

produce a good quality coating, he will experience innumerable paper breaks, wear and tear on the equipment and, an inexscuable amount of machine down time due to maintenance and start-up. In addition to this is the obvious loss of income. Proper alignment is a necessity, but hand in hand with alignment goes proper tension control. If too low a tension is maintained, the paper will drag. If the tension is too high, the edges will tend to curl; both conditions lead to paperbreaks and low operating efficiency.

Skincoat Coating Station Design

Possibly the most important part of transfer coating is the design and operation of the skincoat coating station. The two most common coating techniques used for transfer coating are knife over roll (Figure 4) and reverse roll (Figure 5). However, by far the simplest and most convenient to use is the knife over roll method.

The knife can be of the blunt, radius or J-blade type (Figure 6) and the back up roll is generally rubber or steel. Due to the variations of flow and shear, each knife design is applicable to certain situations; however, the most commonly used knife design is the J-blade, and this is the only one that will be discussed in detail in this article.

The J-blade knife, due to its curvature and extended foot tends to produce good even polymer flow under the knife with a minimum amount of shear. This is very

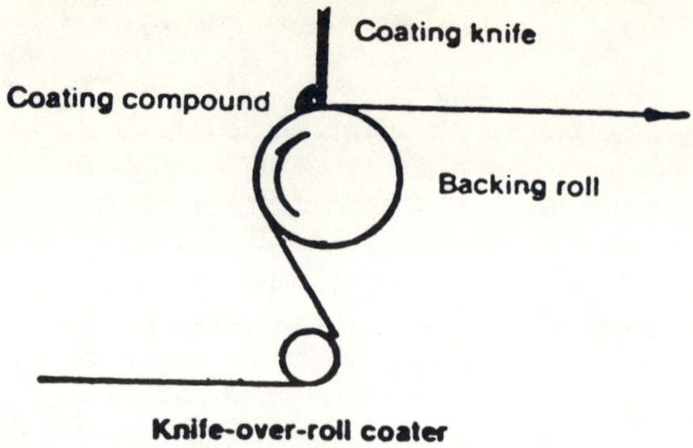

Figure 4. Knife over roll process technique.

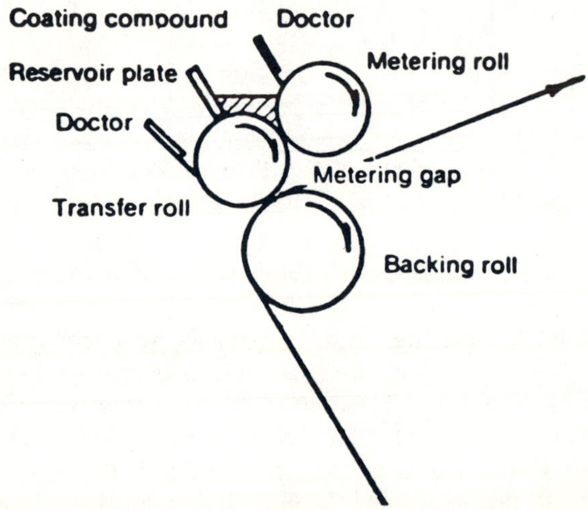

Figure 5. Reverse roll coating.

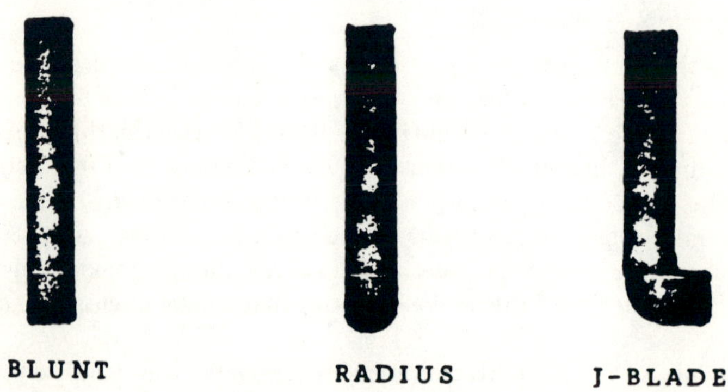

Figure 6. Typical knife blade designs.

important because, as stated earlier, shear has a large influence on the polymer viscosity. In addition to this quality, the J-blade with its extended foot can also be adjusted at different angles until the best polymer flow and application of polymer to the paper are obtained. This is very important, since proper polymer laydown is absolutely necessary for a smooth, uniform skincoat.

Other techniques that aid in the production of a good smooth coating are thorough mixing, with subsequent deaeration and filtration. Filtration can be carried out before or during deaeration. A simple batch method is to filter the urethane through several layers of finely woven nylon cloth and deaerate by allowing the polymer to stand overnight. Just prior to coating, a short mixing time, at low speed, can then be used to blend any particles which have begun to settle out. This technique does not completely deaerate the polymer but does remove the majority of entrapped air.

Another method involves a simple vacuum set up to filter and deaerate simultaneously (Figure 7). This apparatus is simply composed of an open top drum with a detachable filter attached to the cover. The filter plate is covered with several layers of finely woven nylon and the polymer is allowed to filter through the nylon. A small vacuum line can be attached to the top side of the drum and deaeration can be accomplished as the material is filtered. A low vacuum and a somewhat slow filtration rate is used but this method is quite adequate and quite inexpensive to build.

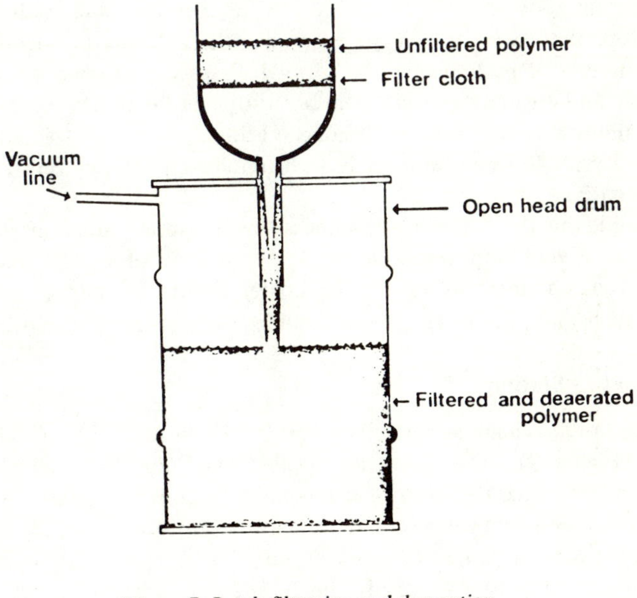

Figure 7. Batch filtration and deaeration.

Skincoat Drying Oven

In addition to deaeration, proper drying of the skincoat is an absolute necessity for the production of a pinhole free urethane film. Figure 8 illustrates a typical three zoned explosion proof skincoat drying train. The drying train should consist of a minimum of two zones and preferably three zones. Each zone should have independent temperature and air controls. Typical drying conditions would be: Zone #1 – 80–120°F., Zone #2 – 120–150°F., Zone #3 – 150–220°F. Air impingement is from the top only in zone #1 and from top and bottom in zones #2 and #3.

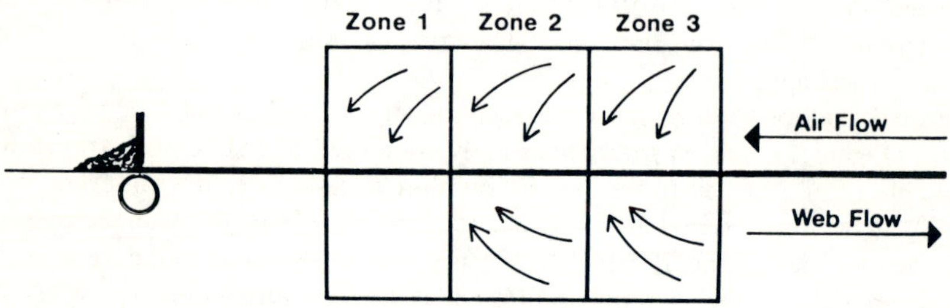

Figure 8. Skincoat drying oven.

By operating zone #1 at a relatively low temperature and with air flow only from the top, the coater is able to evaporate the solvents at such a rate that a minimum number of pinholes will be formed. Passage into zone #2 with its higher temperatures and air impingement from both top and bottom removes the majority of the remaining solvents. Finally in zone #3 the last traces of solvent are removed and the solid pinhole free film is ready to be cooled and carried on to the next step in transfer coating.

Proper adjustment of coating thickness, oven temperatures, air flow and line speed all play a very important part in the formation of a good smooth coating. With these four variables to adjust, the coater should be able to manipulate his equipment such that pinhole free, reproducible coatings are readily produced.

Adhesive Coating Station

A typical polyurethane adhesive is a soft, tacky polymer with a high degree of elongation (Figure 9). It is generally applied via knife over roll coater using a J-blade knife. The quantity applied is dependent upon the solids content, desired adhesion level, composite softness and lamination pressure. Generally speaking, 6–8 wet mils is a sufficient quantity to give adequate adhesion. Almost immediately after the adhesive has been applied, and before it has begun to dry the required fabric is nipped into the adhesive.

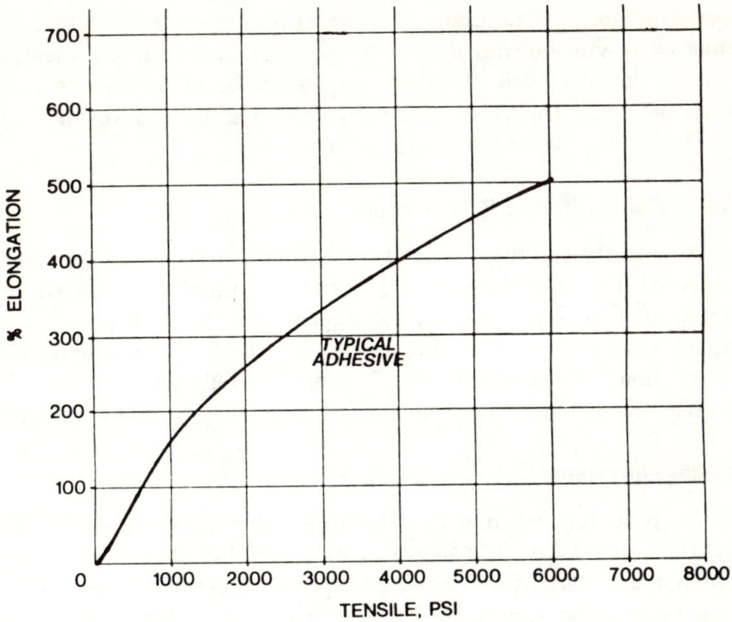

Figure 9. Typical adhesive.

Fabric Lamination

Various designs have been used for controlling the unwind and subsequent lamination of the fabric to the urethane coating. In one method, as the fabric is unwound, it passes over a series of tension control rolls and then a slat roll for control of wrinkles. Finally the fabric passes over another tension roll which brings the fabric in contact with the wet adhesive. This permits the adhesive to wick into the fabric before the final lamination pressure is applied. By varying the contact time between the initial fabric and adhesive contact point and the lamination pressure point, the coater can control the depth of adhesive penetration. This control feature in addition to the lamination pressure control enables the operator to vary the adhesion over quite a wide range. The amount of pressure applied at the rolls is obviously dependent upon the following:

1. Adhesion required by the specific product use
2. Softness of the finished composite
3. Quantity of adhesive applied.

Adhesive Drying Oven

Drying conditions obviously vary according to the solvent system that each adhesive utilizes. However, unless a crosslinking system is beign used, temperatures above 250–275°F are not necessary. The adhesive drying train should have at least a two zone oven with individual zone temperature and air impingment controls. Just as in the skincoat oven, the first zone should be set at a lower temperature than the second zone. This establishes uniform drying and removes the majority of the solvent before the higher temperatures are reached.

Separation of Composite and Release Paper

After thorough drying, the composite and paper carrier are guided over a series of chill rolls. At this point the urethane fabric composite is peeled from the carrier paper and wound up separately for subsequent operations. The release paper is inspected and reused in future coating operation. The finished composite is either sent to inspection for immediate use, or, more generally the case, transported to finishing where the various decorative effects and finally topcoating are applied.

Post Finishing Operations

Decorative procedures such as printing, topcoating and tumbling are then used to convert the coated fabric into an eye appealing product that nearly duplicates its leather counterpart. Various decorative patterns are most often created by the use of the gravure printing technique; however, spraying is also employed in many operations.

The gravure printing technique involves the deposition of a printing ink onto the urethane surface via an engraved cylinder and a doctor blade (Figure 10). Such leather-like looks as kid prints and cloud effects can be accomplished by this technique. Antique patterns and two tone effects are readily obtained by the use of spraying.

A final topcoat can also be applied by either printing or spraying and is generally used to control the final luster, abrasion and slip characteristics of the end product. Today's urethane technology has developed in sophistication such that the coater can incorporate into a finished product a bright, clear or matte appearance with a high or low degree of slip and anything from a dry, waxy, buttery or greasy feel. In fact, today's technology has developed to the point that, in most cases, the coater can produce whatever surface aesthetics the stylist or consumer requires.

In many instances the topcoated products are made even more leather like in appearance by use of a technique borrowed from leather technology. The product is tumbled or milled in a batch, or continuous, tumbler. This permanently installs random creases into the surface of the coated fabric and enhances the appearance of the leather like products. An illustration of the wide variety of physical properties that can be obtained by the transfer coating method is shown in Table 2.

Obviously there are a large number of factors involved in the selection of transfer coating polymers and the techniques necessary to convert them to usable, long lasting products. The more the coater familiarizes himself with the intricacies of these factors the better his products will become.

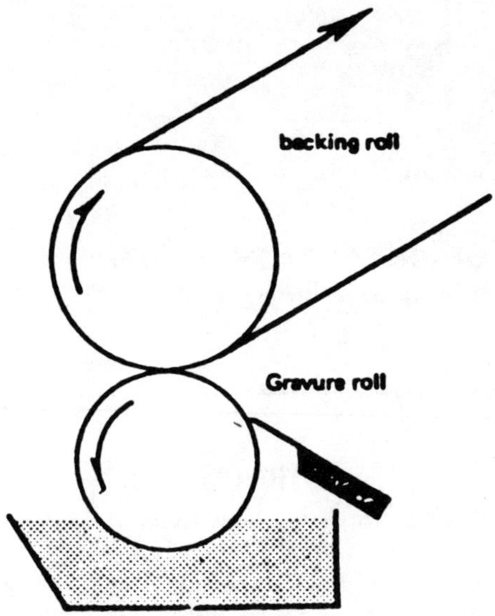

Figure 10. Gravure printing technique.

Table 2. Typical Properties of Transfer Coated Construction

	Footwear	*Apparel and Accessories*
Tabor Abrasion, Cycles[1]		
Initial	200	500[2]
After humid aging	150	300
Balley Flex, Cycles		
Initial	400,000	130,000
After humid aging	250,000	120,000
Bond Strength, pli		
Initial	8	6
After humid aging	6	5

[1] H-22 wheel, 1,000 gram weight
[2] CS 17 wheel, 500 gram weight
[3] Two weeks at 70°C., 95% RH

H. TRÄUBEL

Bayer AG
Dyestuffs Division
Leverkusen, West Germany

THE COATING OF SPLIT LEATHER WITH HIGH-REACTIVE POLYURETHANES

(Received September 25, 1975)

INTRODUCTION

THERE ARE VARIOUS methods available today for coating split leather:

1. Coating (finishing) with polymer dispersions[1] and, if required, subsequent polyurethane finish[2]. Such leathers are suitable for briefcases and for gym and house shoes.
2. The coating of split leather with PVC films for treating by the high-frequency embossing process has recently gained in significance [1]. In this case the coating material consists of an 0.9 mm thick foam layer and an 0.3 mm thick solid top layer. This process has the advantage that one can combine several press-cutting and sewing operations for upper manufacture in one operation. The considerable rationalising effect achieved through this is obtained at the price of a very unleatherlike look since the material is too thick compared with leather.
3. A further method of coating consists of laminating leather with solid polyurethane foils. This process is actually only used at present for industrial goods, since materials coated in this way do not look like leather.
4. Coating with thin polyurethane layers, which can reproduce the grain of the leather in every detail, can be carried out with a special transfer process[3]. Here the polyurethane layers are formed from low molecular compounds on matrices and are applied to leather or other substrates.

The Baycast coating of leather mentioned with high reactive polyurethanes [2] has been carried out in industry for a number of years.

This process follows the known processes of coating split leathers either with compact [3] or microporous polyurethane films [4]. Instead of split leathers, fabric support materials (woven, knitted, bonded-fibre) can be used.

---

[1] for instance with R Eukanol-Resin FD, R Eukanol-Pastes FO + FA
[2] for instance with R Baygen-Products, R Bayderm-Products
[3] for instance with R Baycast Coating

DIRECT COATING WITH POLYURETHANES

Because the technique of coating with polyurethanes is somewhat unusual for leather manufacturers, it will be outlined briefly here:

A NCO prepolymer[4] is mixed using a special apparatus, with the solution of a polyamine[5] and applied by spray. The proportion of the two substances, prepolymer (I) and polyamine (II), are chosen so that they react extremely quickly according to the equation below to become a polyurethane polyurea material (III).

The mixing of the two components (I and II) can, therefore, only be carried out immediately before application:

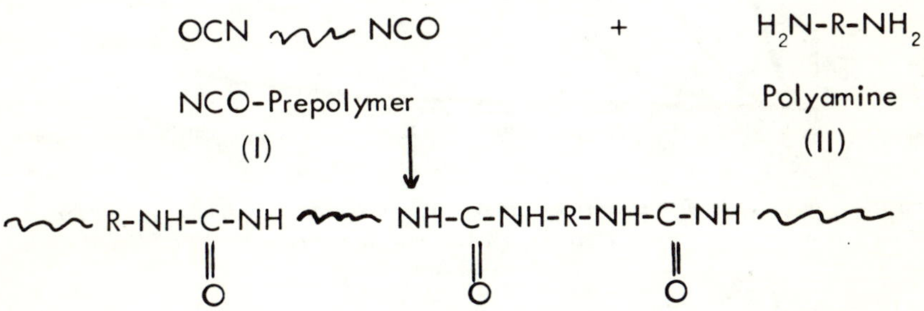

Polyurethane polyurea

(III)

In Figure 1 the course of the reaction with reference to the increase in viscosity of the reaction mixture is shown.

It will be seen from the curve that the products in their normal form cannot be used with conventional machines (spray guns, flow coaters, etc) since their viscosity increases much too rapidly. It can, however, be deducted that thinning the mixture with a highly polar solvent such as dimethyl formamide, slows down the reaction considerably (Curve 4).

The increase in viscosity of, for instance, a 50% solution in dimethyl formamide is slowed down to such an extent that this solution can still be used within about 2 hours at 20°C.

Using a conventional spray gun, which must be cleaned immediately after use with dimethyl formamide, the mixture can be sprayed on to buffed leather; after about 30–40 minutes at 70° in a drying cabinet a coating which may be stacked can

[4] for instance ® Baycast-Adduct LPU
[5] for instance with ® Baycast Hardener HTA

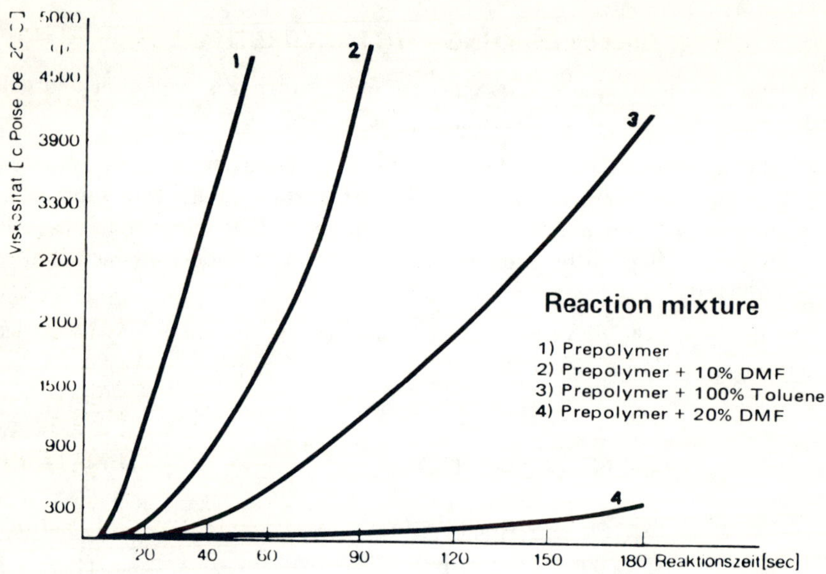

Figure 1. Increase in the viscosity of the reaction mixture (prepolymer and polyamine) at 20° C in solvents (measured using the HAAKE viscometer type VT 180) plotted against the reaction time (secs).

be obtained and this can still be easily embossed after 2–3 weeks. It can be finished further with polyurethane top-coat[6].

These upper leathers have very good wear properties and they can be used for hard-wearing sports shoes, protective work shoes or children's shoes. Proof of their extraordinary wear resistance is, for instance, shown by the abrasion which occurs in the Taber Abrader: after 200 revolutions with a 1 kg load and abrasive disc HS 22 no damage has occurred.

In the Bally Flexometer at least 200,000 flexes in the dry state, 100,000 flexes in the wet state and 10,000 flexes at −25°C have been withstood without damage. Resistance to hot rubbing after 300 revolutions (SATRA, dry felt) and resistance to hot ironing at 150°C are also very good.

TRANSFER COATING WITH POLYURETHANES

The principal application of this process is the coating of split leathers. This coating is mostly applied by means of the transfer process. Here the mixed products are sprayed on to a temporary support, release paper or silicone matrices[7]. The split leather is laid on to the reacting polymer and after 6–8 minutes at about 70°C the leather and PU coating are pulled off the matrix. The coated split leathers can be

[6] for instance with R Desmoderm products
[7] for instance of R Silopren rubber

dressed with polyurethane in a similar way to the direct coated leathers. The same process can also be used for coating fabrics or bonded fibre materials.

If one uses silicone rubber matrices for the temporary support, exact reproduction of the leather surfaces is obtained (Figure 2).

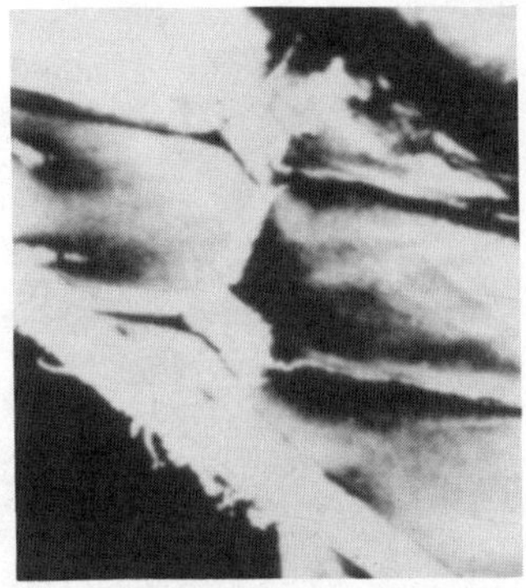

(A) Snake leather

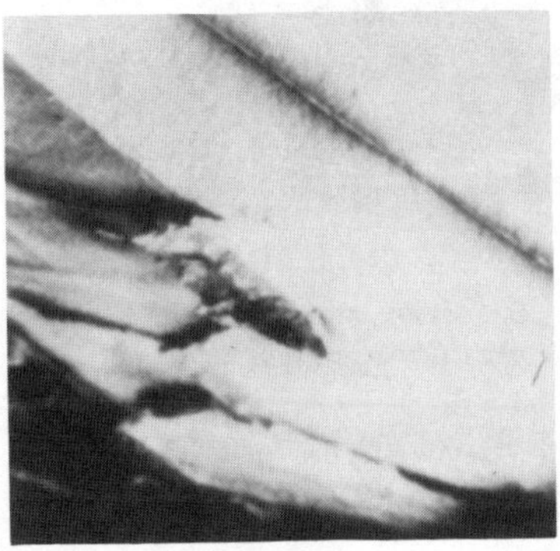

(B) Matrix (negative)

(C) Polyurethane coating

Figure 2. Comparison of the surfaces of the upper leather serving as original (A): the matrices made from it (B): and the coating made from it (C). (Stereoscan photo: magnification 1 cm $\wedge$ 500mm)

The properties of coated split leathers produced in this way are good, as mentioned above. Figure 3 shows children's shoes after a comparative wear test (60 days in each case). One pair was made of corrected grain sides, the other of coated split leather. As can be seen from the illustration, the corrected grain sides have been damaged on the toes during the test period while the coated material is still perfect.

Figure 3. Wear test: Pairs of children's shoes, each worn, alternately, for 60 days. Left: Polyurethane coated split. Right: Corrected grain sides.

THE PROCESSING OF REACTIVE POLYURETHANE COATED SUBSTRATES

During press-cutting the properties of the substrate must be taken into account, depending on whether split leather, bonded fibre fabric, woven material or knitted fabric is used.

If the substrate is split leather, care must be taken that seams at which the material is generally very heavily skived, are not subjected to tension at high loads such as on the toe cap. In skiving, split leather loses so much strength during lasting that the substrate disintegrates. If designs are to be made, in which such a seam lies in a stressed zone, the seam must be reinforced.

When the cut-out part is skived, 0.1–0.4 mm (leather, fabric, bonded-fibre fabric) must remain, so that bonding with hot melt adhesives does not lead to difficulties during the folding process.

After positioning the toe puff (all commercial types are suitable) those types which are usually put in wet after dipping should be well dried (about 20 minutes) before lasting. If the puffs are wet when lasted, slight displacement of the PU film might occur, which could lead to the formation of little pleats at the toe of the upper (vamp). Particularly in the case of freshly coated materials, residual solvent from the puffs might weaken the layer of coating during heat setting.

The lasting or pulling over process can be carried out as for all other shoe upper materials.

The lasting and wiping-in resistance of these materials is good, though it must be assumed that the coating has been carried out properly. No particular precautions are necessary when the lasting margin is roughed. The passage through the drying tunnel or the heat-setter can be carried out in the same way as for all normal upper materials.

As bonding agent for sticking on leather, polyurethane, rubber and PVC shoe soles 2-component polyurethane adhesives are very suitable.

Subsequent finishing with the usual nitro-cellulose products often leads to difficulties on account of poor adhesion. As a shoe finish (depending on the desired degree of gloss) and shadow finishes, those based on non-reactive light-fast polyurethanes[8] in organic solvents are particularly suitable. Damaged shoes can be repaired without difficulty using a repair colour made from such one-component polyurethanes.

In principle, all the usual soling materials can be injected-on. Tests have been carried out with rubber, polyvinyl chloride, polyamide and polyurethane. It was found that the material should preferably be roughed, so that the bond between the injected-on sole and the base material is good.

If the coated materials are finished with 2-component polyurethanes, the bond of the injected-on soles (unless the leather has previously been roughed) is usually not adequate. This finish generally separates so clearly from all soling materials that spew can be removed without damaging the leather.

THE POLYURETHANE REACTIONS AND THEIR INFLUENCE ON THE PROPERTIES

As mentioned previously [2], the properties of the reactive polyurethane layer deteriorate drastically below a certain ratio of the isocyanate groups in the pre-polymer to the amino groups in the polyamine. (NCO/NH_2 ratio < 1.0). From Figure 4 it will be seen that a polyurethane film with good properties always results if more isocyanate groups than amino groups take part in the chemical reaction ($NCO/NH_2 > 1$).

The properties of the film are optimum with a NCO/NH_2 ratio of 1.02–1.4; they are better than all coatings obtained by conventional methods. A considerably higher NCO/NH_2 ratio than 1.5 brings about a considerably longer drying time of the polyurethane layer and hence more difficult manufacturing conditions. When all the isocyanate groups have finally finished reacting with water (moisture from the air), which in this case can take up to 6 weeks, nearly the same tensile and tear resistances are obtained as in the cases of films with a NCO/NH_2 ratio of 1.02–1.4.

Coatings whose NCO/NH_2 ratio is less than 1 cannot be used. The flex resistance,

[8] for instance with R Bayderm Sealer SF
for instance with R Desmoderm-Finish W

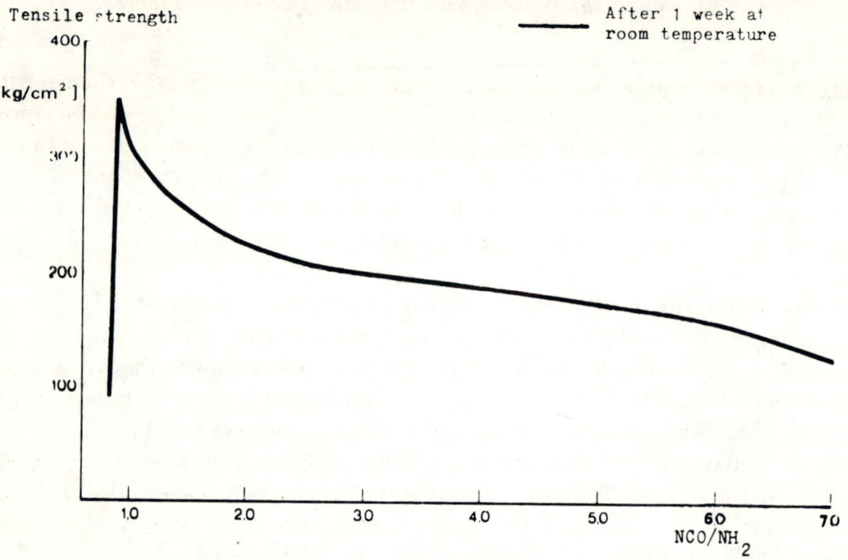

Figure 4(A). Influence of the NCO/NH_2 ratio on the properties of the polyurethane (Tensile strength in kg/cm^2).

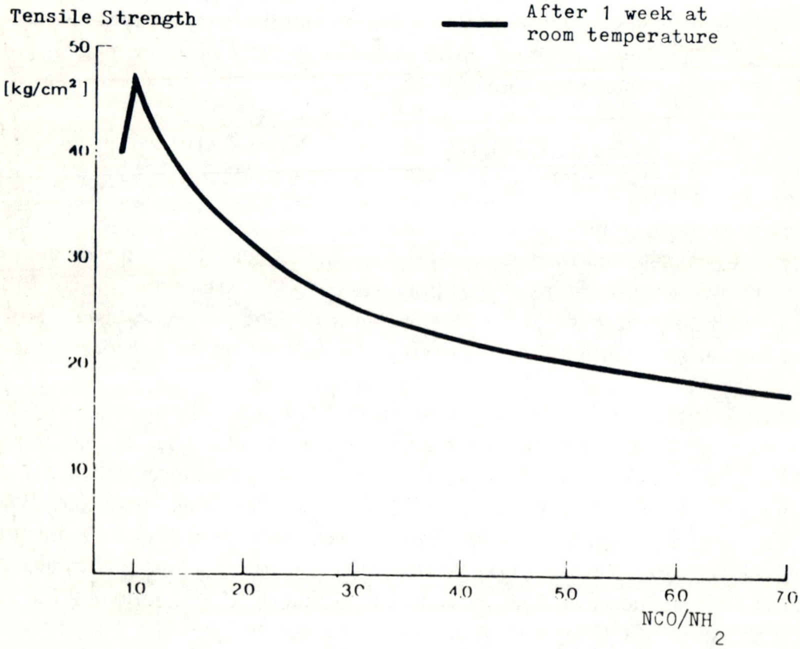

Figure 4(B). Influence of the NCO/NH_2 ratio on the properties of the polyurethane (Tear resistance in kg/cm^2).

high temperature resistance, extension-at-break as well as the solvent resistance are to low; these leathers do not withstand making up into shoes and in consequence lead to complaints.

ECONOMIC CONSIDERATIONS

The economics of the coating methods described here depend on various factors:

1. Production costs:

   Capacity use of the machine, price of raw materials, wages, amortisation of investments, energy and repairs all play the same part here as in other processes in the leather industry. To these must be added product losses if the surface of the temporary support (release paper, silicone matrix) is not used sufficiently by the split leather (cf. Figure 5); in the case of application by machinery (e.g. as opposed to spraying. Translator) the polyurethane is supplied in the form of sheet continuous roll. The leather split often does not cover this polyurethane film adequately.

   Sorting the split leather according to size, and matrix sizes to harmonize with these sizes can minimise these produce losses.

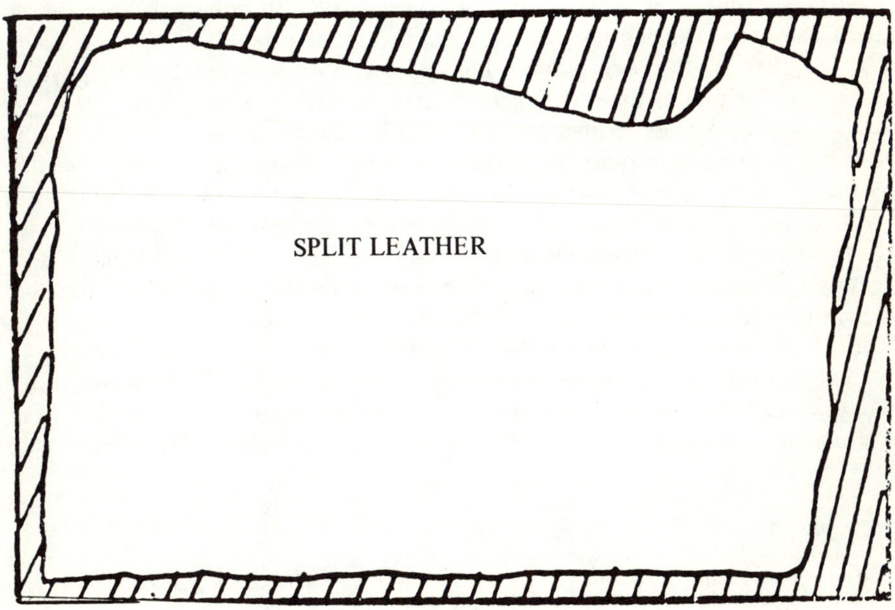

Figure 5. Matrices covered with split leather: the shaded area is polyurethane which is not used for coating (loss).

2. Marketing:

   The economic success of this coating method depends on whether it is possible to capitalize on all the advantages of the goods produced by it. The high abrasion resistance of the products makes them suitable for use in the sports and children's shoe sector as well as for footballs, high-grade brief cases, etc., that is to say articles in which these properties are demanded and paid for. Since with the aid of the matrix technique one can follow fashion trends very quickly, it is possible to copy particularly "in" leather surfaces at short notice (snake leather, reptiles etc.).

3. Relation between the prices of split leathers and the sale price of the coated articles:

   The price of the raw material (split leather) depends on whether there is a suede fashion or not. The price of the split leather is, incidentally, the greatest factor in the calculation. As a rule of thumb it can, however, be said that with a production of 3–4 hours per day, the process brings in a profit if the finished coated leathers attains 1.5 times the selling price of normal pigmented splits.

CONCLUSIONS

A survey about the reactive coating process [5] with polyurethane has been given. The properties and the appearance of the coated products are extraordinary, since one can in this way produce high-grade leather with outstanding physical properties. The making up of the materials into shoes has been discussed.

The coating process treated here in detail can enable leather factories to considerably broaden their range of products and open up areas in which split leather by itself is only partially suitable. Since the coating itself is very uniform and it is possible to have identical surface patterns on various substrates using the matrices, one can by means of this technique very easily produce what might be called "set" coatings. This method of coating can therefore be a suitable "entree" for a leather firm into textile coating.

In the final discussions about the economics of this process the dependence of success or failure on certain parameters is shown. Whether one wishes to invest money in such a technique, buy a suitable machine and operate such a development is a commercial decision which, in the opinion of the author, this article should help in reaching.

REFERENCES

1. Anon. Schuhtechnik 66 No. 1 (1972), pp. 12–14.
2. H. Träubel, JSLTC, Vol. 57 (1973), pp. 59–62.
3. H. Herfeld and J. Steinlein, Das Leder *24* (1973), pp. 98–104, 118–124.
4. B. Zorn, Das Leder *22* (1971), pp. 147–150.
5. H. Träubel, Das Leder, September 1974, pp. 161–166.

ANTHONY J. DURANTE, PH.D.

Union Carbide Corporation
Tarrytown, NY 10591

AND

BARRY ZIMMERMAN, PH.D.

Union Carbide Corporation
New York, NY 10017

MECHANICALLY FROTHED URETHANE FOAM IN SIMULATED LEATHER COMPOSITES

A Review of Coated Fabric Technology and the Requirements for Foam Composites

(Received October 6, 1975)

INTRODUCTION

THIS PAPER PRESENTS a review of the current technology of fabrics coated with both vinyls and urethanes, with particular emphasis on the latter in producing simulated products for shoe uppers, apparel, upholstery and handbags. The history of the development of urethane coated fabrics and current market size and potential are also included.

It has been demonstrated that frothed foam may be used in place of the pile of napped and sheared fabrics to separate skin coat from fabric. Fabric composites containing frothed foam rival the esthetics of leather, and can be produced with mechanical frothing machines and slightly modified but conventional fabric coating equipment. The manufacturing details for the production of coated fabric composites having frothed foam as the inner core have been developed on semi-commercial equipment where (a) the complete composite was assembled continuously and (b) fabric was attached directly to pre-gelled foam *or* to the dried foam by means of a separate adhesive.

The frothed urethane process for depositing continuous, uniform castings of controlled density foam lends itself to current fabric coating technology, and has excellent potential by offering leather-like products on inexpensive unnapped backing fabrics.

BACKGROUND

Fabrics coated with natural gums and resins were used as mummy wrappings to preserve the regal dead of ancient Egypt. Other natural coating materials such as linseed oil and rubber latex have been used over the years to produce oil cloths, sun shades, and raincoats. Fabrics coated with early synthetics such as the cellulose derivatives led to coated fabrics which gave leather its first real challenge. As the demand for coated fabrics increased over the years, especially in automotive interiors and convertible car tops, more flexible and durable products were required. The newer polymeric film formers such as PVA and PVC, developed during World War II, found their way into coated fabrics for rainwear, tarpaulins, ponchos, tenting, floats, and life jackets. Today fabric coating employs a variety of methods and principally involves two basic polymers: poly(vinyl chloride) and polyurethane.

PVC, the most widely consumed fabric coating material, has experienced tremendous growth over the past ten years. By 1970, the volume of PVC used in the United States for fabric coating exceeded 375 million pounds for about 500 million square yards of coated fabric [1]. At about 8 yards per car, automotive markets claimed about a third of the U.S. production. Depending on several quality control and formulation parameters, vinyl coated fabrics have been made to suit diverse end-uses, functional and decorative. Until the early 60's, it had replaced leather in all but a very few applications where the natural esthetics, strength, and breathability of leather were necessary. Expanded vinyl on soft knitted fabric combined the toughness of vinyl and the suppleness of leather. Despite all their advantages, vinyl coated fabrics suffered from problems such as cracking, development of hard and soft spots as a result of preferential plasticizer migration, staining, softening, and the lack of breathability.

In the early 60's, polyurethane fabric coatings began replacing vinyl in the more exotic and luxurious end-uses where vinyls would not fit. Urethane polymers used as quality floor surfaces and marine finishes have been used as early as 1940, and coating of fabrics with these polymers was a logical extension of this growing technology. Urethanes are inherently tougher than vinyls, and 1–3 mils can often replace 15–20 mils of vinyl in most coated fabric applications [1].

The two major methods used in coating other polymerics onto fabric have been applied to urethanes. The methods are *direct coating* and *transfer coating*. The first involves casting of viscous polymers directly onto a web of fabric to produce primarily functional materials such as ponchos, raingear, ski jackets and parkas, tenting, and the like. The second method is used to produce leather-like products which have rivaled the esthetics of leather because of their luxurious feel, dry hand and good drape, and especially because of their light weight and breathability. Transfer coating depends on coating release paper, which may be plain or embossed, with a thin film of solution cast thermoplastic polyurethane. The coated paper passes through a drying oven in which solvent from the skin coat is carefully removed. A second coat of the same or a different polymer is applied, and then fabric is laminated into the wet polymer. The complete assembly is dried in a second oven after which the release paper and coated fabric are separated. This booklet will be limited to leather substitutes made by the *transfer coating method.*

GROWTH AND MARKETS

Much of the success of the urethane simulated leathers can be attributed to the combination of excellent physical properties, and the loft and body of the napped fabrics onto which they are customarily coated. The first of these unusual products arrived in the U.S. from Spain and Italy in 1962. Their rich leather-like appearance had tremendous consumer appeal, but as the fads and fashions of the apparel and shoe industries would have it, quantity soon outranked quality. Finished goods were produced and released to consumers who soon complained of delamination, chipping, enbrittlement, wrinkling, poor washability and dry cleanability, coating defects, and lack of color stability. Therefore, to best serve this growing market and to insure its future growth, major changes in urethane resin technology and fabrication techniques were initiated. Thus, the market was slow in developing, and not until the late 60's were suitable improvements made.

Today, urethane coated fabrics are enjoying excellent growth, largely because of urethanes' exceptional chemical and physical properties. They are intrinsically tougher than all the other synthetics competing in this marketplace; they are abrasion resistant, elastic at low and high temperatures, water and gas permeable, chemically inert, and most important, they create the leather esthetics no other polymer can. Unheard of 15 years ago, urethane coated fabrics grew to a 50-million linear yard domestic market in 1973 divided among some 35 coaters. Nearly double that production is expected for 1976. Although growth is anticipated in all areas, some inroads into automotive are expected in 1977.

The premium ends of the urethane coated fabric market are shoe and upholstery applications where physical and chemical properties of the composite must be at their maximum to pass the performance requirements. For example, typical properties for unsupported skins are:

Tensile	5000-7000 psi.
Tear	400-700 p/i
Elongation	400-600 %
Skin-fabric adhesion	>7 p/i

Many synthetics and leather will survive the flex and abrasion test conditions, but the costs per square foot to the shoe trade vary widely.

CLOTH SUBSTRATES

The substrates commonly used for coated fabrics are usually the industrial variety of fabrics and are selected by end-use. In the case of coating with polyurethane, fabrics are limited to those which can be uniformly napped and sheared. These are usually cotton or more recently cotton/polyester blends of the types listed in Table 1. For transfer coating of thin films of urethane, the raised pile of the napped and sheared fabrics serves to eliminate mark-off of the weave pattern onto the thin urethane skin, and to cushion and separate skin from fabric. The density of the nap and the height of the resulting sheared pile determine the softness, drape and body of the coated product. In short, the success or failure of the final product is largely determined by the quality of the napped and sheared substrate [3].

Table 1. Substrates for Urethane Coated Products

Fabric Type	Width, inches	Weight, yds./lb.	Ends x Picks
Sateen	57	1.18	124 x 64
Sateen	58	1.21	96 x 64
Soft Filling Sateen	62	1.60	38 x 46

The napping and shearing operation is tedious and costly, adding about 25 percent to the cost of the finished fabric. Domestically, this operation is at best a "bottleneck," and the resulting textiles suffer from wide variability in the critical areas of nap density and pile height. Excellent quality is attained in Europe where processing, testing, and machinery have been developed to satisfy the standards of the huge, established coated fabrics industry, and where these fabrics had previously been optimized for suedette clothing materials.

An idealized cross-section of a typical urethane coated, napped and sheared fabric is illustrated in Figure 1 [4]. Optimum quality is achieved when the pile is uniformly imbedded into the tie coat – excessive penetration beyond the tie coat and into the skin coat yields potential failure sites, and a stiff, "boardy" product; insufficient penetration yields poor adhesion, a loose fitting skin ("floating") and non-leatherlike esthetics. Despite the very strict process requirements, some very excellent simulated leathers have been produced using this technology [5].

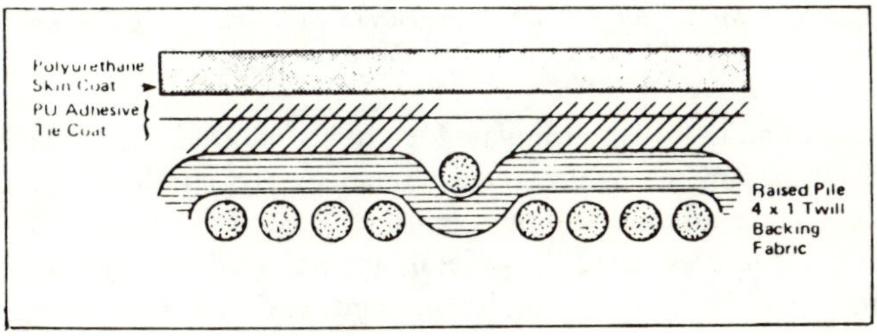

Figure 1. Idealized structure of polyurethane coated fabric [Ref. 4].

FOAM AS THE INNER CORE OF FABRIC COMPOSITE

A very desirable method of separating skin from fabric in much the same way as they are separated by the pile of a napped substrate is with a thin layer of a soft, flexible foam. This can be achieved either by laminating foamed sheeting between skin and fabric, or by casting foam or foam ingredients onto skin, followed by gelation and fabric lamination. Most attempts at achieving this have all fallen short of complete success for reasons such as irreproducible, unmanageable processing requirements, short pot life, poor control of foam thickness and density, poor adhesion of foam to both skin and fabric, along with poor leather esthetics and high cost. Only mechanically frothed urethane foam has successfully overcome many of the shortcomings of other systems. Frothed poly(vinyl chloride) has been used in this application, but composites suffer from the attendant shortcomings of vinyl itself mentioned earlier.

Frothed urethane foam as the inner core of a coated fabric may be used as the adhesive between skin and foam. Alternately, the foam may be cast and cured on the skin, and then adhered to fabric with a separate adhesive or tie-coat. The former is illustrated in Figure 2, where foam ideally separates fabric from skin to prevent weave mark-off, adds loft and body to the construction, and provides an excellent leather break and feel. The table top foam foundation for skin leads to high abrasion and snag resistance.

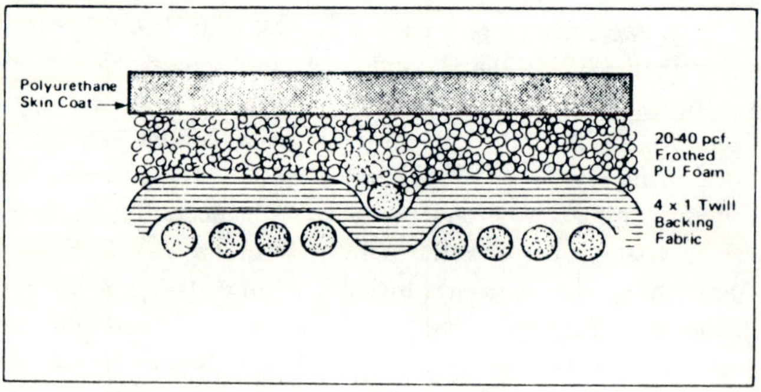

Figure 2. Idealized structure of polyurethane/fabric composite having a frothed urethane core.

The proposed end-use of the simulated leather product dictates the construction and composition parameters which must be met by the combined skin, foam, and fabric. In Table 2 are assembled some typical constructions which have been found appealing for the end-uses listed. In general, apparel grade goods must be washable and dry-cleanable, as well as supple, drapey, and light-fast; upholstery grades must be highly resistant to hydrolysis, discoloration, and abrasion; shoe grades must be extremely durable and resistant to flex cracking, abrasion, and moisture; handbag

Table 2. Typical Construction Parameters by End-Use

End-Use	Surface	100% Modulus, psi.	Prime Character	Density, pcf.	Thickness, mils	Hand	Type
Apparel	draggy	600-800	colorfast and cleanable	20-30	5-15	soft, limp	tricot, knit soft sateen
Upholstery	dry	1,000-1,500	hydrolytic and abrasion resistant	25-35	15-25	pliable	knit, sateen
Shoes and boots	smooth and dry	800-1,000	excellent abrasion and flex life	40	20	stiff	soft sateen, nonwoven
Handbags and luggage	various	400-800	scuff resistant	20-30	5-15	firm	drill, sheeting

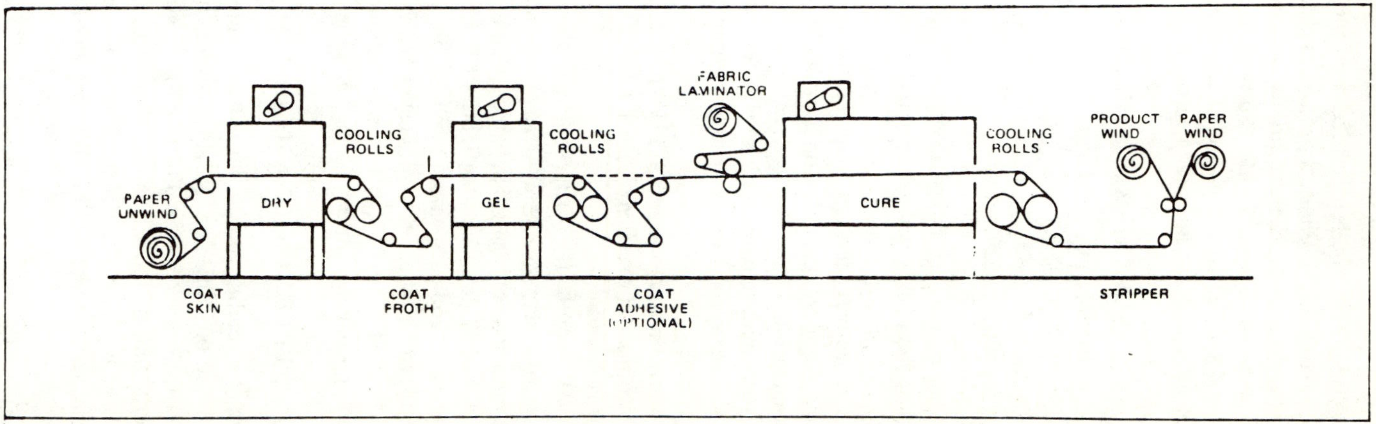

Figure 3. Typical casting line for frothed foam composites.

and luggage grades, the least critical applications of all, must merely be resistant to scuffing. Assuming all physical properties can be met by the proper combination of construction variables, the end products' appeal can be greatly influenced by color, hand and drape. The value for 100 percent modulus listed in Table 2 is largely a measure of the feel to be expected from the skin of a finished composite. Slight changes in the skin modulus can be detected readily by touch.

ADVANTAGES OF FROTHED URETHANE FOAM

Mechanically frothed urethane foam as the inner core of fabric composites has successfully overcome the shortcomings of other systems, and has provided the first really excellent and potentially commercial composites which contain foam. Reasons for this are:

(a) Conventional frothing machines (Oakes, SKG, Euromatic) used for latex or vinyl may be readily adapted to meter-feed and dispense urethane foam intermediates.

(b) The froth is relatively stable at room temperature, but gels to a tacky stage accepting fabric within 50 seconds at 300°F, and cures tack-free in 3–4 minutes at that temperature.

(c) When spread coated with a knife coater, control of gauge is excellent.

(d) Conventional fabric coating equipment can be utilized with little or no modification.

(e) Adhesion to fabric and to PU skin coat is excellent.

(f) Various and excellent leather esthetics can be quickly achieved by adjusting foam density and thickness.

(g) The smooth, flat surface of the foam provides an excellent foundation for thin urethane skin coats and leads to increased abrasion resistance and permits ready production of high gloss, patent products.

(h) A multitude of backing fabrics can be included in the composite for diversified product end-uses without relying on the quality and availability of conventional napped and sheared fabrics listed in Table 1.

(i) The cost of the foam for many applications offsets that of the processing involved in producing quality napped and sheared textiles.

PROCESSING OF FROTHED FOAM IN FABRIC COATING

Simulated leather composites which employ frothed foam as the inner core may be produced on conventional fabric coating lines. A schematic diagram of a typical three-station casting line is illustrated in Figure 3. This arrangement adapts itself to the production of vinyl, foamed vinyl, and conventional urethane coated fabrics. The procedure for making polyurethane foamed composites have been worked out on a pilot casting line at S. D. Warren, Westbrook, Maine, and can be described as follows:

(a) At the first station, urethane skin is cast at 6–8 wet mils onto either flat or

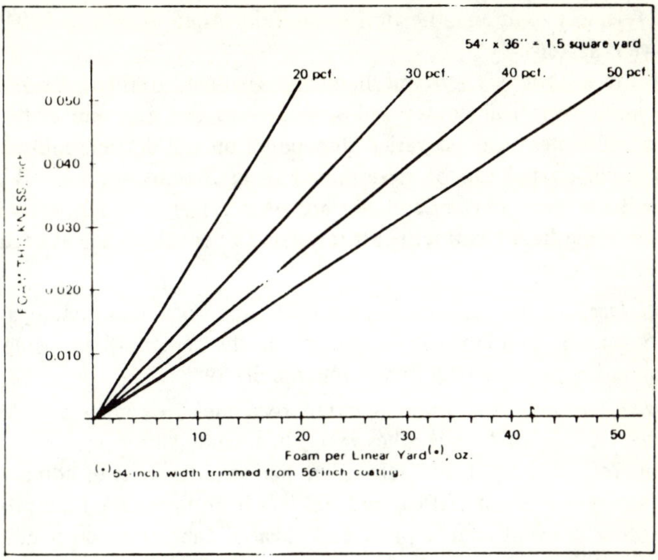

Figure 4A. Add-on weight per linear yard (*) vs foam thickness and density.

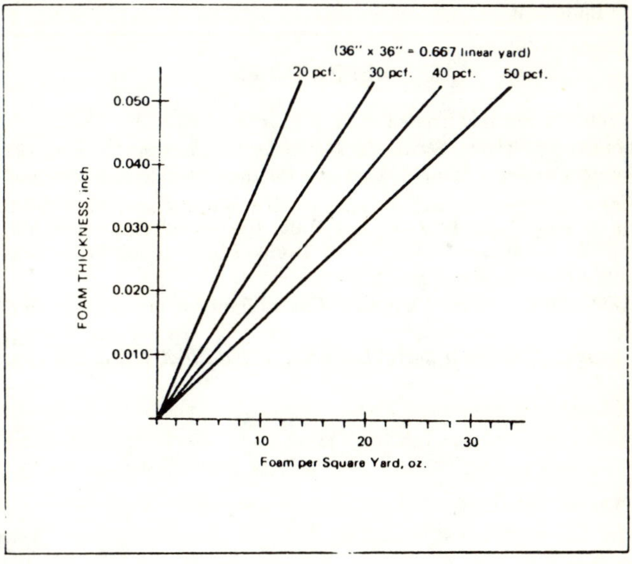

Figure 4B. Add-on weight per square yard vs foam thickness and density.

embossed release paper, and the film is carefully dried at 200 to 300°F depending on the solvent system.

(b) At the second station, mechanically frothed urethane foam is deposited behind a knife over roll coater and spread onto the dry skin coat. The foam's density and thickness can be varied depending on the desired end-use. The exact weight of foam coated can be determined from Figures 4A and 4B which relate weight per linear yard and per square yard, respectively, to density and thickness.

(c) After some heat treatment in the second oven, the foam is ready to accept fabric by either:

(1) *Pre-gel lamination* – foam is partially cured to a tacky state at 300°F for 50–55 seconds, and fabric is deposited into the "green" foam using a minimum of laminating pressure (0.5 lb per linear inch), or

(2) *Adhesive lamination* – foam is cured to a tack-free state at 300°F for 3–4 minutes. A solvent based adhesive is then coated at 6 wet mils at the third station, followed by direct fabric lamination into the wet adhesive. The laminating pressure is not critical and has less bearing on the final product's hand than when conventional napped and sheared fabrics are applied to urethane films directly.

(d) Depending on the lamination technique, the third oven can either continue the cure of the pre-gelled foam or dry the adhesive.

(e) After a final cooling operation, the finished product is stripped from the release paper and spooled separately. The release paper may be reused several times depending on handling.

REFERENCES

1. Bernard Taub, "Urethane Coated Fabrics for Automotive Upholstery Applications," *Urethanes in Coated Fabrics*, Technomic Publishing Co., Westport, Conn., 1973.
2. R. T. McSweeney and R. J. Athey, "Urethane Coatings for Textile Substrates", *Urethanes in Coated Fabrics, op cit.*
3. F. J. Beaulieu and M. D. Troxler, "Substrates for Coated Apparel Applications", paper presented at the American Association of Textile Chemists and Colorists' Symposium on Coated Fabrics, Newton, Mass., March, 1973.
4. D. Popplewell and L. G. Hole, "Urethane Coated Fabrics" in *Urethanes in Coated Fabrics, op cit.*
5. Ernest R. Kaswell, *Wellington Sears Handbook of Industrial Textiles,* Wellington Sears Co., New York, 1963.

RICHARD F. HOLDEN

Fleming Joffe Jentra, Ltd.
311 West State St.
Johnstown, New Jersey 12095

FINISHING OF POROMERIC POLYURETHANES

(Received November 13, 1975)

INTRODUCTION

POLYMERIC SHOE UPPER materials are most commonly vinyl laminated or vinyl coated woven fabrics or even, in a few instances, nonwoven fabrics. Urethane coated fabrics are in this same category. The term polymeric mistakenly infers physical properties and qualities normally associated with the more costly poromerics.

The term "poromeric" was coined by E. I. DuPont to describe their shoe upper material, Corfam®. Experts have been writing definitions for the word ever since. The common stipulations, found in an assortment of definitions, are: similarity to leather in structure, strength, thickness; handle and temper; and the property of moisture permeability. Structurally, the poromeric upper materials currently on the market vary widely. Tanera® at one extreme is an almost 100% nonwoven textile. At the opposite extreme lies Porvair® which is 100% cellular urethane. The balance of the commercial products, with which the author is faimilar, lie in between and consist of saturated nonwoven substrates with cellular urethane grain layers in varying proportions. All of these products, however, exhibit the common property of moisture vapor transmission. This is a measurable property and is normally expressed as weight per unit area per time, i.e., grams per square meter per 24 hours. Whatever the units or test method, the standard of comparison is leather and a permeability equal to leather should be the ultimate objective. All poromerics, regardless of structure, also seek to approach leather in their other basic properties: tensile strength, elongation, thickness, temper and moisture absorbancy. All of the properties so far discussed are design-engineered into a basic greige goods which we as finishers use as raw material. Our job is to turn these greige goods into a merchandisable shoe upper material having the color, texture, gloss and particular appearance wanted by the shoe manufacturer while preserving the basic properties built into the greige goods.

*Presented at the Urethanes in Footwear Seminar, sponsored by Technomic Publishing Co., Inc.

FINISH REQUIREMENTS

Apart from style, the primary requirement of finishes is that they should be durable, so that the attractive appearance of the new shoe is retained as long as possible. The finish should be soil-resistant and easily cleaned without suffering from the cleaning process. It should not discolor or loose gloss nor should matte finishes become glossy. Finishes must be repellant to water and should not blotch, stain, discolor or swell when brought into contact with water. Finishes must adhere firmly and should not crack or peel. They must be elastic enough to withstand folding along the edge of a shoe pattern yet be hard enough to withstand scuffing and rubbing off. Finally, they should not become tacky when exposed to heat or brittle in extreme cold and they should be waterproof but not impair the breathability of the substrate.

All of the above are met on leather every day using ordinary leather finishes applied by well-known, relatively unsophisticated, equipment in much the same way as has been done for the past fifty years. Why then should the finishing of a poromeric substrate present more than a few mechanical challenges associated with continuous web handling? The answer is that, if we were to take Jentra® greige goods or any other poromeric for that matter and process it through a conventional leather finishing plant in the normal manner, the end product would be a stiff, cardboardy material with poor tear strength, poor finish adhesion and generally unsatisfactory physical properties. Conversely, if we were to process a side of chrome leather through our plant in our normal manner, the net result would be equally unsatisfactory.

FINISHING AGENTS

Much credit is due DuPont for developing the early chemistry and technology for finishing poromerics. In fact, until approximately five years ago, all of the poromeric finishing in the United States was done using essentially the same methods as DuPont used in the finishing of Corfam. The process consists of base coloring with a pigmented self-curing acrylic emulsion polymer, embossing at high temperature and low pressure and top-coating by spray or gravure printing with a plasticized solution of cellulose acetate butyrate perhaps containing some transparent pigment or dye.

Crosslinking acrylic emulsions were used because of their high flexibility, good moisture vapor transmission, ease of pigmentation, resistance to re-wetting after curing, and their generally good adhesion to the microporous urethane grain layer.

Cellulose acetate butyrate lacquers were used primarily for their high abrasion resistance and their excellent resistance to discoloration from both heat and ultraviolet light. Crease dulling and the tendency to become brittle in other than very thin coatings are the major drawbacks of these cab coatings.

In truth, the same systems used to finish Corfam® years ago will still produce finished Jentra® capable of passing nearly all of our quality control specifications.

Were it not for the continuing improvement of flexible urethane finishes and the availability of non-discoloring aliphatic urethanes at reasonable prices, I suspect we would still be finishing "Jentra®" as we finished "Corfam®" in those early days.

Rapid growth, in recent years, of the urethane-coated fabric market has created a sizable market for flexible urethane coatings and has spurred urethane research and development to a pace unequalled since the late 1950's and early 1960's. The net result is the availability of urethane coating systems which meet virtually all of the requirements of the poromeric finisher.

In addition to meeting certain basic physical requirements, specific quality assurance tests must also be met. These specifications are largely the direct result of our failure experience of the past and the recent development by major shoe manufacturers of product approval specifications which must be met before a poromeric upper material can be considered for use in shoes. Of these numerous specifications, the only ones discussed here are those which are controlled in the main by the finishing process.

TESTING POROMERIC URETHANES

The first and most important of these is flex-cracking. Bally flex is still the test most frequently used in the United States and is the test which we use at Fleming Joffe Jentra. All new finishes are flex tested dry to failure or one million cycles. Samples are also flexed after immersion in water at room temperature for three hours. Wet flex tests are run for 50,000 cycles and then re-wet for one hour before continuing another 50,000 cycles. This process is repeated up to 500,000 cycles or failure. A third sample is aged for seven days at 180°F and 95% relative humidity and then Bally flexed the same as the dry Bally sample. No more than fifty percent loss of flex life is allowed after humid aging.

A system of ranking failure is used for Jentra® and is more easily shown than discussed. Basically the type and extent of cracking is ranked from one through five with one being perfect and five being a failure. More significance is attached to sharp, deep cracking than fine checkering and any type of leafing or finish loss is ranked worse than simple cracking. Samples are normally run until they reach a level of five, recording the number of cycles flexed, but can alternatively be run to a given number of cycles and then comparatively ranked against other samples at the same number of flex cycles.

Abrasion resistance is evaluated by edge wear testing. This test consists of rubbing the sample, mounted on a ½-inch mandrel under four pounds load, against a stationary strip of number ten cotton duck. Samples are run both dry and after immersion for three hours in water at room temperature. Dry samples are run 1,500 cycles and wet samples 50 cycles and then rated visually for finish removal and crocking using a system similar to that described for flex cracking. No substantial finish removal or crocking is allowed.

Fadeometer tests are run for 100 hours for color stability and continued to

1,000 hours for evaluation of UV degradation. This is a visual evaluation for crazing, checking, gloss loss, etc. Adhesion is checked by crosshatching the finish and attempting to remove it with high tenacity transparent tape and by immersion for three hours in boiling water and then abrading the finish with a fingernail.

In addition to these physical specifications, a candidate finish must be capable of producing the aesthetic properties required by the stylist and the mechanical properties necessary for the type of applications being evaluated. For aesthetic properties, the main requirement is that suitable colorants, both pigments and dyes, be commercially available at reasonable cost. In addition, gloss control and control of surface feel are important in top finishes.

The application requirements are largely dictated by the specific coating application being evaluated. An ideal coating for gravure application may not be usable for silk screen printing, dip coating or spraying even if solvated to the appropriate viscosity and evaporation rate. If a candidate coating meets the physical and aesthetic requirements, it can usually be formulated to meet our application drying and curing conditions if the supplier is willing to modify the backbone polymer solution to do so.

If one surveyed all of the commercial U.S. poromerics, it is believed that the coating techniques which have been used or are used now cover all of the possible means of applying finishes. Transfer coating techniques were used for smooth patent finishing of Aztran®, and Corfam® with the process being much as urethane coated fabrics are now made. The main difference was that the urethane skin was cast and dryed on release paper and then transferred by heat reactivation as needed.

Curtain or flow coating has also been used for patent finishing or poromerics, particularly for fancy patent effects. In this case, only piece goods are handled and the process is identical to patent leather finishing.

It should be noted, however, that high gloss patent finishes can only be achieved at the expense of moisture vapor transmission. This is also true of genuine patent leather and the method of application chosen will have a negligible effect on this property. Foot comfort in patent finished poromerics must depend on absorption of moisture by the non-woven substrate or shoe lining and subsequent release of this moisture by the pumping action of the foot in the shoe.

Base color coatings have been applied by conventional air spray, airless spray and meyer rod coating. The choice is largely a matter of economics. Dip coating and meyer rod coating present none of the finish waste problems of spray application and require no provision for the handling of airborne particular matter.

At Fleming Joffe Jentra, we also have the capability of silk screen printing on poromerics. This process is used where multi-color indexed printing is desired as we do not have the necessary equipment for indexed printing by gravure.

All other coating applications are by gravure printing whether for imparting design or as protective or gloss control coats. This is by far the most cost efficient coating method which we use, both from the material consumption and time/labor points of view. It is also the means by which substantial style can be built into the product without building up so much coating that moisture vapor transmission is unduly retarded.

EMBOSSING TECHNIQUES

One of the most important operations in the finishing of a poromeric shoe upper material is embossing. Excessive compaction of the saturated non-woven backer will substantially reduce MVT. Fusion and collapse by heat and pressure of the microporous urethane structure will have the same effect. On the other hand, sufficient heat must be used at embossing to overcome the plastic memory of the microporous urethane or the embossed pattern will rebound as soon as the finished product is subjected to heat in the shoe manufacturing process.

Consequently, the method used for embossing is unique for poromerics and is more like molding than conventional types of embossing. The poromeric surface is first preheated and is then passed between heated nip embossing rolls. Minimum embossing temperatures vary with individual poromeric manufacturers but are in the range of 320° to 350°F. Pressure requirements at the nip are low to avoid compaction and the lineal speed is substantially slower than other shoe materials.

At one time or another, poromerics have been finished with just about every classical type of urethane. Two package aromatic urethanes are still used for patent finishing. These crosslinking systems give the highest initial gloss of any type which we have evaluated. In addition, gloss retention, scuff resistance and solvent resistance are excellent. One package blocked systems can be used in our equipment but have not performed as well for us as two package systems. Two package aliphatic systems are still extremely high in cost and not as readily available as the aromatic types.

Moisture cured one package systems, while giving at their best competitive properties to the two package aromatics, have a tendency to gas during curing and bubbles are formed in the finish. In addition, reliability of curing is totally unsatisfactory and flex crack resistance is not as good as with two package aromatics.

URETHANE ELASTOMERS

The most versatile systems are the urethane elastomers. These polymers are now available from a number of suppliers with physical properties ranging from sticky adhesives to very hard lacquers. Within the same range of physical properties utilized by urethane coated fabric manufacturers, there are polymers suitable for just about everything from basecoating to top finish coating. The softer polymers are good for intermediate print coats and the harder polymers within this range are excellent protective topcoats.

Pigmentation of urethane elastomers is relatively easy and commercial dispersions are readily available. The IPA/TOL or IPA/TOL/MEK solvent systems are excellent choices for dye solubility. In addition, a wide variety of additives are now available for gloss control and surface feel. One can have just about any kind of aesthetic properties he could ask for.

One major new development of these polymers is the ready commercial availability of non-discoloring aliphatic systems at virtually no extra cost. This has removed one of the last barriers to their widespread use in poromeric finishing. The extremely high flex crack resistance, toughness and abrasion resistance and hydrolytic stability of these polymers assures them of continued use for a long time.

EMULSIONS

Urethane emulsions, while not new to the industry, have not kept pace with other urethane systems. We have evaluated a number of emulsions and find most to have the same group of problems. Very few urethane emulsions are non-ionic or alkaline in pH. Consequently, they are difficult to compound and color. Leveling and wetting are poor and their chemical and mechanical stability leave much to be desired. Lastly, emulsion costs have stayed very high. Nevertheless, given the steady rise in solvent prices and increasingly stringent solvent emission controls, it would seem these should be the systems of the future.

The alternative to emulsion systems would be 100% solids systems. These are discussed from time to time by suppliers but are still in the future. The logic of minimizing solvent use and eliminating the energy requirements necessary to evaporate solvents, or even water for that matter, make 100% solids systems extremely attractive to the poromeric finisher or to any coater for that matter.

To summarize, urethane finishes of various types have become the most common raw materials of the poromeric finisher. Much work is required, however, if poromerics and other urethane-consuming shoe upper materials are to retain viable market positions.

We cannot continue to use three pounds of petroleum-derived solvent to utilize one pound of urethane and we need to reduce our energy requirements by eliminating evaporation processes before we are forced to do so by unsympathetic government intervention.